AMERICAN CICHLIDS II: LARGE CICHLIDS

Wolfgang Staeck · Horst Linke

Wolfgang Staeck · Horst Linke

American Cichlids II

Large
Cichlids

A Handbook for their
Identification, Care,
and Breeding

The American Cichlids are presented
in 2 volumes:
Cichlids I: Small Cichlids
Cichlids II: Large Cichlids

© **1985 Tetra-Verlag**
Tetra-Werke Dr. rer. nat Ulrich Baensch GmbH
D-49304 Melle, P.O. Box 1580, Germany

All rights reserved, incl. film, broadcasting,
television as well as the reprinting

1st completely revised edition (1995)

Distributed in U.S.A. by
Tetra Second nature (Division of Warner Lambert)
3001 Commerce Street
Blacksburg, VA 24060

Distributed in UK by
Tetra Sales, Lambert Court,
Chestnut Avenue, Eastleigh Hampshire S05 3ZQ

Typesetting: Fotosatz Hoffmann, Ruppichteroth
Lithos: Flotho Reprotechnik, Osnabrück
Translation from German:
Herprint International cc, South Africa

Final Text Editing: Dr. Wolfgang Staeck

Printed by Mandarin Offset, Hong Kong

ISBN 1-56465-169-X

WL Code 16758

Contents

FOREWORD

This book is a translation of a completely revised, updated, and expanded edition of the volume on large American Cichlids that appeared in 1985. A comparison with the original edition shows that our understanding of the cladistic relationships and evolution of the Cichlids native to the Americas has meanwhile drastically changed. Obviously this had to have an effect on their systematics and taxonomy, so that many species today have different scientific names. This is particularly true for the Central American Cichlids where systematics has considerably changed due to the new concepts applied to the genus *Cichlasoma* within the past decade. The South American Cichlids were also subject to major revisions resulting in taxonomic changes. One consequence of these works was the definition of half a dozen new genera for large South American Cichlids and the revalidation of ten generic synonyms. The enormous increase of our knowledge is not only shown by the number of newly discovered species or those investigated taxonomically for the first time ever, but also by the vast number of species imported for the first time. The appendix "Index of Synonyms" lists scientific names that used to be in common use among aquarists during the last decades and their modern equivalents. This list enables you to find all species in the main part of this book without problems.

Our sincere thanks are due to Sven O. KULLANDER (Naturhistorika Riksmuseet, Stockholm) who in some instances assisted with the difficult identification of fishes, Heinrich MAULHARDT, Leonidas Yarhua TARIEUARIMA, and Erich HNILICKA, who advised and assisted us in Peru and Mexico respectively, and Ingo SCHINDLER and Lothar SEEGERS who during our travels were patient and reliable companions even when it became tough.

We also like to extend our appreciation to Jens GOTTWALD, Ingemann HANSEN, Lutz KRANEFELD, Erwin SCHRAML, Ernst SOSNA, and Uwe WERNER for providing us with photographs and information.

Berlin, 1995   Wolfgang Staeck   Horst Linke

AMERICAN CICHLIDS

Recent estimates indicate that far more than a thousand different Cichlids live on earth, many of which have never been scientifically described. To date, nearly 500 species have become known from the Americas alone. Cichlids are basically tropical fish which do not tolerate temperatures below 20 °C for a longer period of time. Water-temperature thus seems to be the factor limiting the distribution of this family. This fact explains why Cichlids generally occur between the Tropic of Cancer and the Tropic of Capricorn and show the widest diversity in regions adjacent to the equator. They are therefore common in central and northern South America and in Central America, but are not found in North America.

To date, some 35 genera containing about 400 species are known from South America (KULLANDER 1994) and some 16 genera in Central America accommodate nearly 100 species. Among the Central American Cichlids there are about a dozen species whose distribution is restricted to northern Mexico and which may therefore be referred to as North American Cichlids. One species, *Herichthys cyanoguttatus*, even occurs in southern Texas in the US. The northernmost localities of this species

Vieja bifasciata provides an idea why Cichlids enjoy such a popularity as aquarium fishes worldwide.

lie in the drainages of the Rio Pecos and Rio Nueces and form the northern limits of Cichlids in the Americas.

In the south, we find a similar situation with *Crenicichla lacustris* (comp. KULLANDER 1981) occurring in regions farther south than any other Cichlid. This species was recorded from Patagonia in the vicinity of the village of Puerto Madryn between 42 and 43° S. This apparently represents the southern distributional limit of American Cichlids.

Out of the almost 500 American species some 100 may be considered Dwarf-cichlids which were dealt with in the first of the two volumes on American Cichlids. The others are large fishes which are focused on in this book. This type of splitting induced the necessity to choose different concepts for the two volumes on American Cichlids. The limited number of species in the first volume made it possible to portray the small Cichlids almost completely, whereas this volume can only take a comparatively small selection of large species into consideration. Here the following criteria were applied. First, such Cichlids should be portrayed that are particularly interesting and recommendable for the aquarium. Nevertheless, a representative overview should be provided on the diversity of forms and ecologies that are found among large American Cichlids. Moreover it was undertaken to also include a reasonable number of species that made a first appearance in the aquarium hobby within the last years and about which little information has so far been published. The limits were obviously set by the volume of the book.

American Cichlids have been known to Ichthyology for a surprisingly long time. For example, *Cichla ocellaris* was described by BLOCH & SCHNEIDER as early as in 1801, and in 1840 *Johann Natterer's neue Flussfische Brasiliens nach den Beobachtungen und Mittheilungen des Entdeckers beschrieben* by Jacob HECKEL, ichthyologist at the Museum

CENTRAL AMERICA

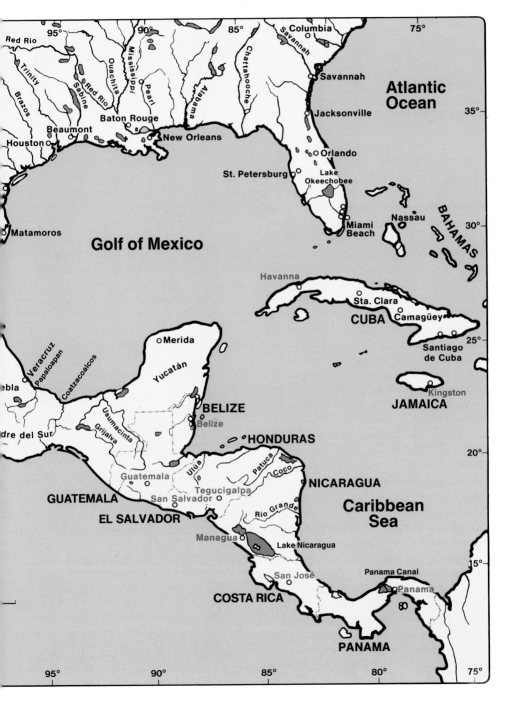

95°   90°   85°   Columbia   75°
Red Rio   Savannah
Mississippi
Trinity   Ouachita   Chattahoochee   Savannah
Ouachita   Red Rio   Pearl   Alabama   Atlantic
Brazos   Sabine   Alabama   Ocean
Baton Rouge   Jacksonville   35°
Beaumont   New Orleans
Houston   Orlando
St. Petersburg   Lake
Okeechobee
Nassau   BAHAMAS
Matamoros   30°
Golf of Mexico   Miami
Beach
Havanna   Sta. Clara
CUBA   Camagüey
Merida   25°
Veracruz   Santiago
Papaloapan   de Cuba
Coatzacoalcos   Yucatán
ebla   Kingston
Usumacinta   BELIZE   JAMAICA
Grijalva   Belize
dre del Sur   HONDURAS   20°
Uliua   Patuca
Coco
Guatemala   NICARAGUA
Tegucigalpa
GUATEMALA   San Salvador   Caribbean
EL SALVADOR   Rio Grande   Sea
Managua   Lake Nicaragua
San José   Panama Canal   15°
COSTA RICA   Panama
80
PANAMA
95°   90°   85°   80°   75°

für Naturgeschichte in Vienna, already provided a representative selection of American Cichlids. A number of original descriptions were published by the ichthyologist STEINDACHNER, also of Vienna, whose publications on the "Chromiden Mexikos und Zentralamerikas" appeared in 1864, followed by "Die Süsswasserfische des südöstlichen Brasilien" in 1874, and "Chromiden des Amazonasstromes" one year later.

Two more classical standard references in ichthyology that enhanced the knowledge on American Cichlids were published by GÜNTHER who worked at the British Museum of Natural History in London. In 1862 his *Catalogue of the Fishes in the British Museum* appeared, followed by *An account of the fishes of the states of Central America, based on collections made Capt. I.M. Dow, F. Godman, Esq., and O. Salvin, Esq.* in 1869. The year 1904 saw the publication of an important study of the biology and taxonomy of Cichlids by the Frenchman PELLEGRIN that dealt with many American species.

A revision of all American Cichlids that simultaneously was a kind of survey of all knowledge gathered up to this point was published during the years 1905-1906 by the London based ichthyologist REGAN. In 1908 he added an annotated list of all Central American Cichlids known by then.

From then on, the most important results of the ongoing research in American Cichlids were no longer published in Europe but in America. This development had started as early as in 1898 when a standard reference on North and Central American fishes authored by JORDAN & EVERMANN appeared. During the years 1906 to 1913 it was MEEK who, on the basis of his various collecting trips to Mexico and other Central American countries, published his studies on the Cichlids of this region. HASEMANN wrote *An annotated catalogue of the Cichlid fishes collected by the Expedition of the Carnegie Museum to Central South*

America, 1907-1910 in 1911. Other American ichthyologists who worked with South American Cichlids before the First World War were EIGENMANN and FOWLER. HUBBS then published the results of expeditions to Central America during the 1930s in which Cichlids were particularly emphasized. At the same time, AHL, who worked at the Zoologisches Museum in Berlin, described several new species of Cichlid fishes.

After the Second World War, especially MILLER and BUSSING continued the tradition of American ichthyologists by conducting research in Central America and, partially in co-operation with other authors, describing newly discovered Cichlids.

In Europe, the Belgian ichthyologist GOSSE, who in the company of King Leopold of Belgium had undertaken several collecting trips to Central and South America, focused on American Cichlids. The year 1975 saw his revision of the genus *Geophagus*. During the 1960s and 70s LÜLING of the

Table 1:

Example of black water in Peru

pH		6.0
Conductivity	µS	17
TH	°dH	0.12
Ca	°dH	0.07
Mg	°dH	0.05
CH	°dH	0
Na	mg/l	1.9
K	mg/l	1.1
NO_3	mg/l	1.6
PO_4	mg/l	<0.1
Cl	mg/l	<7
SO_4	mg/l	n.d.
Zn	µg/l	5
Pb	µg/l	n.d.
Cd	µg/l	n.d.
Cu	µg/l	7
CO_2	mg/l	1

This water sample was taken by the authors in a creek near Jenaro Herrera in the drainage system of the lower Ucayali in July 1983 and analysed by Dr. G. Ritter in the laboratory of the company Tetra.

The Brazilian Rio Negro is a typical black water river. Its water is clear, but has the colour of tea.

Museum Alexander Koenig in Bonn repeatedly visited South America and brought home, new Cichlids as well as information about their biology. Since the late 70s the Swedish ichthyologist KULLANDER has extensively worked on South American Cichlid fishes. Besides a considerable number of original descriptions he published revisions of various genera, among which there was a fundamental revision of the South American representatives of the genus *Cichlasoma* in 1983. This work is of particular importance as it results in drastic reductions of the huge genera *Cichlasoma* and *Aequidens*. Most of the Cichlids previously assigned to these systematic units were transferred to other genera and thus have now different name combinations.

In contrast to African Cichlids which are predominantly mouthbrooders, the American counterparts in this family spawn on certain media. Among the about 100 Central American Cichlids of the tribe Cichlasomini there is not even a single mouthbrooder. The 60 odd mouthbrooders in South America belong to the genera *Gymnogeophagus*, *Geophagus*, *Satanoperca*, *Aequidens*, *Bujurquina*, and *Heros*. Except a very few species, e.g. *"Geophagus" pellegrini* and *steindachneri*, which belong to the specialized ovophile mouthbrooders where the eggs are taken into the mouth immediately after they have been laid and where the female cares for the brood, all other American mouthbrooders are larvophile. They spawn in the fashion of openbrooders, form a parental family-structure during the incubation period of the eggs, and only become mouthbrooders once the larvae have hatched.

The majority of large American Cichlids are typical openbrooders. But there is also a small group of species that show a clear trend towards breeding in hidden places. This minority includes the South American species of *Crenicichla* and representatives of the genus *Archocentrus* from Central America which show a well developed sexual dimorphism and dichromatism and assume sex-dependent roles during periods of parental care. Only very few large Cichlids, such as *Neetroplus nematopus* and *Acarichthys heckelii*, may be termed cavebrooders. This contrasts with the majority of dwarf-cichlids which are typical cavebrooders.

Most of the large American Cichlids are robust and resistant fish. Due to their surprising ability to adapt they inhabit a variety of water-bodies which not only enormously differ with regard to size, flow, hardness and pH, but also in the structure of the bottom, exposure to light, temperature, availability of food and hiding-places, as well as clarity. In the Amazon river system three different types of water are distinguished based on the characteristic degree of clarity and colour of the water. They are typical of specific geological, geomorphic, and climatic conditions. So-called black water rivers, such as the Rio Negro and the Rio Cururu, carry clear, transparent water that is dark brown like strong tea. This colour is produced by large amounts of rotting leaves and other plant material carried into the rivers during the rainy season when wide areas of the rainforest are flooded.

The second category includes the so-called clear water rivers, examples being the Rio Xingu and the Rio Tapajoz. They are characterized by very clean, transparent water that is poor in sediments and shows green to yellowish green colours. The third group contains the so-called white water rivers, such as the Rio Solimoes and the Ucayali, which carry large amounts of inorganic material, are therefore turbid, and have the colour of loam. Visibility here is often less than a few centimetres. These three types of water are of course not always found in their pure qualities, but intergrade and mix with one another in any form imaginable.

The Central American Fish regions
(derived from MEEK, 1904, MILLER, 1966, and BUSSING, 1976)

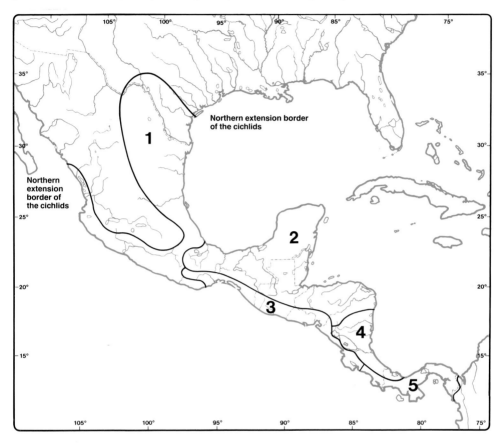

1: Rio Grande-Panuco-Balsas-region (about 20 Cichlids)
2: Rio Papaloapan-Grijalva-Usumacinta-region (about 50 Cichlids)
3: Chiapas-Nicaragua-region (about 15 Cichlids)
4: Rio San Juan-region (about 15 Cichlids)
5: Isthmian region (about 25 Cichlids)

The Central American Cichlids are limited to five fish regions and were originally assigned to the four genera *Herotilapia* PELLEGRIN, 1904, *Neetroplus* GÜNTHER, 1869, *Petenia* GÜNTHER, 1862, and *Cichlasoma* SWAINSON, 1839. Since its revision (KULLANDER 1983), *Cichlasoma* is restricted to only a dozen South American species with the Central American Cichlids being spread over the eleven genera *Amphilophus*, *Archocentrus*, *Chuco*, *Copora*, *Herichthys*, *Paraneetroplus*, *Parapetenia*, *Theraps*, *Thorichthys*, *Tomocichla*, and *Vieja*. The South American Cichlid genera are only represented in the Isthmian fish region, i.e. in the very south, by one species of both *Aequidens* and *Geophagus*. The northernmost region that is typically exemplifed by the genus *Herichthys*, is considered to belong to North America by some authors.

15

Clear water rivers like the Rio Arruya (Peru) carry very clean water that is usually of a greenish colour.

White water rivers like the Rio Ucayali (Peru) are very turbid from organic drift material and mostly have a loam-yellow colour.

For the waters of the Amazon it is generally true that their pH-values range from slightly to clearly acidic and that they are extremely poor in dissolved minerals so that their total and carbonate hardnesses are very low. These attributes apply to black

water in particular which can cause Cichlids from this type of water to be quite demanding in captivity and very difficult to breed.

Although the three types of water described afore are particularly characteristic of the Amazon region in Brazil, they can

Table 2:
Example of clear water in Peru

pH		7.2
Conductivity	μS	142
TH	°dH	4.9
Ca	°dH	3.6
Mg	°dH	1.3
CH	°dH	4.7
Na	mg/l	3.9
K	mg/l	0.5
NO_3	mg/l	0.8
PO_4	mg/l	<0.1
Cl	mg/l	<5
SO_4	mg/l	n.d.
Zn	μg/l	11
Pb	μg/l	n.d.
Cd	μg/l	n.d.
Cu	μg/l	27
CO_2	mg/l	10

This water sample was taken by the authors from the Rio Chinipo in the drainage system of the upper Ucayali in July 1983 and analysed by Dr. G. Ritter in the laboratory of the company Tetra.

Table 3:
Example of white water in Peru

pH		7.1
Conductivity	μS	154
TH	°dH	2.9
Ca	°dH	2.9
Mg	°dH	0
CH	°dH	3.9
Na	mg/l	9.0
K	mg/l	1.8
NO_3	mg/l	0.8
PO_4	mg/l	<0.7
Cl	mg/l	<5
SO_4	mg/l	n.d.
Zn	μg/l	7
Pb	μg/l	n.d.
Cd	μg/l	n.d.
Cu	μg/l	13
CO_2	mg/l	10

This water sample was taken by the authors from the Paca Cocha near Pucallpa in the drainage system of the central Ucayali in July 1983 and analysed by Dr. G. Ritter in the laboratory of the company Tetra.

also be found in slightly different forms throughout tropical South and Central America. For example, there are typical clear and white water rivers in Peru, Bolivia, and Mexico, but they usually do not carry extremely soft water. Instead, their water shows values of moderate hardness, and for some clear water rivers in Mexico that drain the Sierra Madre Occidental, even a very high hardness is typical. It is rather unusual to find very low, acidic pH-values in water-bodies of these countries. Here, the concentration of hydrogen-ions more typically ranges between 7 and 8. In contrast to clear and white waters, real black water is only exceptionally found outside the drainage systems of the rivers Amazon and Orinoco. The still widely distributed opinion that all South American Cichlids would originate from very soft and acidic waters is therefore incorrect.

Except a few specialized species which prey upon fish and hunt in open waters like *Cichla ocellaris,* American Cichlids generally inhabit the zones near the embankments of their home waters where they find sufficient cover and places to hide. This has partly to do with their breeding biology, as they only find media suitable for spawning in these sections.

Rocks will neither be found in white water rivers with a muddy bottom nor in black waters that usually have a sandy ground. The only spawning medium available to the Cichlids of these waters is wood from the surrounding forest, of which, however, there is usually plenty. Logs, branches, and roots also provide the necessary hiding-places for the fish. Clear water rivers in contrast often have a rocky or stony bottom that offers the fish ample facilities to retreat. It is common for many stagnant water-bodies in the tropics that they are densely covered with floating plants like Water hyacinths *(Eichhornia crassipes, E. azurea)* and Water lettuce *(Pistia stratiotes).* They also provide cover to the fish and are

Typical clear water rivers can also be found in Central America (Rio Chamula, Chiapas, Mexico).

therefore favourite places for Cichlids. It is less commonly known that the often 15-20 cm thick layer of rotting leaves on the bottom of the water-bodies is a favourite residence of certain Cichlids. The inhabitants of these biotopes include *Apistogramma*-species and semiadult *Crenicichla.*

Table 4:
Example of clear water in Mexico

pH		7.68
Conductivity	µS	1200
TH	°dH	15.2
Ca	°dH	7.6
Mg	°dH	7.6
CH	°dH	12.8
Na	mg/l	0.7
K	mg/l	0.7
NO_3	mg/l	< 5
PO_4	mg/l	n. d.
Cl	mg/l	< 5
SO_4	mg/l	34
Zn	µg/l	< 0.1
Pb	µg/l	0.2
Cd	µg/l	< 0.1
Cu	µg/l	0.4
CO_2	mg/l	8

This water sample was taken by the authors from the Rio Nututun near Palenque (Chiapas, Mexico) in April 1983 and analysed by Dr. G. Ritter in the laboratory of the company Tetra.

The trees show that the Amazonian rainforest is flooded for metres during the rainy season.

Certain species of *Bujurquina* from the clear water rivers of Peru even use submerged leaves as substrates for spawning. When threatened they may pick up the leaf with their mouth and flee to another place.

It is surprising to note that the aquarium literature has paid so little attention to the fact that the water-bodies in the Amazon region are exposed to considerable fluctuations with regard to the depth of water, even more so as this condition has wide ranging effects on the biology of the Cichlids and any other fish. According to the seasonal rhythm the water levels regularly rise and fall by several metres between the rainy and the dry season. A difference of six to eight metres is therefore not uncommon for many water-bodies in the Amazon region.

Observations in the aquarium revealed that some South American Cichlids (e.g. *Uaru amphiacanthoides*) are voracious vegetarians, and others (e.g. *Mesonauta festivus*

and *Pterophyllum scalare)* prefer to spawn on plants and not on rocks near the bottom. At the first glance, this seems to be in conflict with the fact that the natural habitats of these Cichlids lack aquatic vegetation altogether except for a few floating plants on the water surface. This controversy is, however, only a seemingly one. There are many indications that the Cichlids of those South American waters exposed to explicit seasonal changes have relatively sharp defined breeding seasons that correspond with the yearly high waters (comp. STAECK 1982). During these periods the water-bodies far surmount their banks and set the surrounding rainforest for several metres under water. The water level can be so high that bushes and the lower branches of trees are deep under the water surface. It is therefore obvious that it is not aquatic plants that play an important role in the lives of these Cichlids, but submerged parts of land vegetation that serve as food resources and spawning media.

Certain peculiarities in the reproductive biology of some Cichlids from white or black water rivers can also be explained by circumstances in their natural habitats. The species of *Mesonauta* and *Symphysodon* never spawn on the bottom, but at some distance from the ground whereby they prefer vertical surfaces rather than horizontal ones. Their larvae are not deposited in pits. The parents use the sticky excrete of their head glands to attach them to branches, roots, or floating plants near the water surface. By living in white or black water, these Cichlids are forced to move their breeding activities from the bottom to the surface because it is only there that they find an adequate visibility.

All the aforesaid shows that some behavioural patterns which may seem unusual in the aquarium serve a biological purpose. All of the sudden they make sense when there is an occasion to observe the fish in their natural environment.

THE SUITABLE AQUARIUM
FOR LARGE AMERICAN CICHLIDS

The majority of large American Cichlids are robust and hardy fish, attributes that apparently make them recommendable for the aquarium. This is particularly true for the Central American genus *Archocentrus* which exclusively incorporates moderately large species. It also offers several Cichlids which will breed without effort under almost every kind of care and are therefore also suitable for aquarists with limited experience in the care of fish. Organisms that are quite tolerant of a wide range of environmental conditions are termed euryoecious by ecologists, i.e. those who investigate the interaction between organisms and their environment.

On the other hand, there are a variety of stenoecious species among the larger American Cichlids, i.e. those with a very limited tolerance to variations in environmental conditions. They will only feel comfortable when almost all environmental factors agree with the conditions under which they live in nature. These are the so-called problem fish which are difficult to keep in an aquarium and even more difficult to breed. Most Cichlids belonging to this group come from the black water rivers of the Amazon basin. They include species where even experienced breeders in many attempts failed to prompt them to breed.

The long-term keeping and breeding of Cichlids can only be successful under the precondition that the setup of the aquarium corresponds as closely as possible with the specific biology and environmental requirements of the respective species. This inflicts that before you start to prepare a tank for the keeping of large American Cichlids, you should first gather all relevant information on how they live and reproduce in their natural biotopes. It is only the exact knowledge of all environmental factors that have an impact on the life of the Cichlids in nature that enables you to more or less exactly emulate these conditions in the aquarium. In this connection it is important to know about natural water-temperatures, light, flow, type of bottom, structure and texture of banks, as well as chemical and physical particulars of the water, especially its total and carbonate hardness and pH.

Even today there are still regions in the Amazonian rainforest that are largely inaccessible and uninvestigated and therefore hold species of fish unknown to man.

19

Besides factors originating from the unanimated environment, there is also an influence of the animated surrounding that determines the possible existence of fish. It is, for example, very important to know which algae and higher plants and other animals occur in a particular water-body as all organisms may have a serious impact on the biotope they share with the Cichlids, be it in the form of a resource of food, a competitor, or a potential predator.

It is often overlooked that the setup and decoration of a tank contain consequential environmental factors that may either positively or negatively influence the thriving of a fish. Occasionally, Cichlids disappoint their keepers by being drab and not showing the bright colours that are expected. A reason then might be that they originate from intensely sunlit, clear water rivers, but are kept in an aquarium where bottom and decoration are very dark. They have in such case just adapted their colouration to this surrounding. Other species may in contrast look very pale and uncoloured because the bottom of the tank is covered with white gravel although they originate from shaded rainforest creeks with a dark brown to blackish ground of rotting leaves.

As is to be expected, large Cichlids require very spacious tanks. Although there are quite a variety of medium-sized species, e.g. in the genera *Biotodoma, Bujurquina, Mesonauta, Thorichthys,* and *Archocentrus,* which can be kept and bred in aquaria with a length under one metre, even in their cases larger tanks are far more recommendable. Only if sufficient space is provided, they will they show their interesting behaviours to the full extent. The precondition for a suitable environment is therefore a tank of approximately one and a half metres in length.

It may sound paradox, but the care of semiadult Cichlids requires a larger aquarium than that necessary for housing a compatible pair of fully grown adults. This has to do with the pronounced intraspecific aggression that most specimens show towards conspecifics beyond a direct partnership relation. It is complemented by the fact, that the majority of large American Cichlids become relatively calm, little active fish. In this connection it is a basic rule to disturb the fish as little as one possibly can which includes to never take their eggs, larvae, or fry away before their natural caring behaviour has ceased. An interference like this can easily trigger disputes and fights even among well-established partners that, under the restricted available space of an aquarium, may result in the death of one fish.

A recommendable procedure to reduce the intraspecific aggression is to house not only one species in a tank, but several different ones. Obviously it is then necessary to choose an even larger aquarium than would be necessary to accommodate a single pair as every species should be in a position to happily live in a territory of its own. This method simultaneously provides the ground for the most interesting observations and creates a basis for behavioural studies.

Suitable co-inhabitants of such community aquaria are also found among some other families of fish. Representatives of the family Loricariidae, the South American Sucker Catfishes, such as the genera *Glyptoperichthys, Peckoltia, Hypostomus, Ancistrus,* and *Hemiancistrus* have proven to be compatible with Cichlids. Certain Characins are also adequate under the condition that they are of approximately the same size. They are particularly recommendable in the case of very skittish Cichlids that are kept on their own because the active swimming of the Characins helps them to overcome their shyness. Central American Cichlids are perfect in the company of Live-bearing Toothcarps.

First and foremost, the natural decoration of an aquarium for American Cichlids involves that there are an appropriate number of hiding-places and sanctuaries. In their white or black home waters, these fish find shelter among submerged logs and branches, in clear water rivers also between and under rocks. Tropical lakes and ponds usually have a cover of floating plants which provides additional protection. A well-conceptualized arrangement of sight barriers and hiding-places in the tank can also simplify the organization of territorial borders.

The topic of whether or not an aquarium for large Cichlids must be greened is subject of extraordinarily controversial discussions among aquarists. Typical Amazonian white and black water rivers are generally bare of higher aquatic plants rooted in their bottoms, the reason being that the heavy turbidity or brown colouration of these waters does not allow the necessary light to reach the bottom. The extreme changes of water levels in South American water bodies are also an obstacle difficult to overcome for aquatic plants. Purists therefore favour the idea that plants are inopportune in an emulated biotope like this. On the other hand there can be no doubt that the greening of an aquarium may greatly enhance its aesthetic appeal. It must also not be ignored that individual groups of plants can well contribute to the comfort of the fish since they provide cover.

In general there is little that can be said against having plants in an aquarium as long as the vegetation is not in opposition to the life requirements of the fish. In order to assure that there is adequate swimming space left, an aquarium for large American Cichlids should only be planted sparsely. This attitude is supported by another point.

The widespread, though unjustified opinion that Cichlids are opposed to plants is not based on them being vegetarians; species with such feeding ecology clearly are a minority among American Cichlids. When plants are destroyed by the fish, it usually has to do with their breeding behaviour and an inadequate decoration of the

Vieja guttulata is one of the Central American species of Cichlid with a highly variable colour pattern.

aquarium. It is part of the instinctive behavioural patterns of large American Cichlids to clear the immediate vicinity of their clutches or pits in which the larvae are kept of any obstacles that might prevent them from discovering an approaching predator early enough to effectively defend their offspring. For this reason they tend to remove all plants from the centre of their breeding territory. It is therefore advisable to use plants rather sparsely and strategically in an aquarium for large American Cichlids.

Recommendable plants for such tanks are single large specimens of *Echinodorus*, species of *Anubias,* and the fern *Microsorum pteropus,* which should be attached to roots or rocks near the water surface where it anchors itself after some time.

Thoroughly washed coarse sand or fine gravel should be used for the bottom of an aquarium for large American Cichlids. Grain-size should not be too large otherwise there is a risk that the larvae which are

deposited in pits by these openbrooders slip between the individual grains and will be lost. In the cases of *Geophagus* and *Satanoperca,* sand is a must as these fish will only feel at ease when they are enabled to satisfy their natural instinct of chewing through the substrate in search of food. Cichlids from white and black water rivers generally feel comfortable above a dark coloured bottom, while those from clear waters usually prefer light surroundings.

As to how far the keeping of Cichlids is constantly successful and whether breeding attempts have positive results largely depends on the physical and chemical attributes of the water in the aquarium. If not explicitly mentioned otherwise, the Cichlids portrayed in the following should be kept at water-temperatures between 25 and 27 °C. For breeding it might be advisable to increase the standard value by two degrees Celsius since this speeds up the development of eggs and larvae and thus

Large Cichlids like *Vieja synspilum* require hiding places between roots and rocks to feel comfortable.

considerably shortens the period of time between spawning and free moving fry.

A particular property of South American clear and black water rivers is their cleanliness. White water rivers carry large amounts of inorganic material along, but even they are usually not polluted with organic metabolism products. As a result of their large food requirements large Cichlids have a significant turnover of materials so that an effective, powerful filter is recommended to mechanically keep the water of the aquarium clean. Even more important are regular partial exchanges of the water to remove invisible, harmful endproducts of the fish's metabolism. Such intensive care for the water is especially important for many Cichlids originating from Amazon black waters as they are most susceptible to increased levels of nitrite and/or nitrate in their environment.

It is still a widespread opinion among aquarists that American Cichlids require particularly soft water with high acidity. Numerous analyses of the water in the natural habitats have meanwhile revealed that this is wrong. To a certain extent, it may be true for Amazon Cichlids from the river systems of the Orinoco, Rio Negro, and Rio Solimoes that indeed include species that will only breed in such water. For Mexican and other Central American fish as well as those from the central and upper regions of the Ucayali in Peru it is inapplicable as they are usually found in relatively hard water with a pH between 7 and 8. As the individual requirements of those species can vary considerably with regard to the chemical properties of the water, the information on natural habitats and tables showing the results of water analyses should be given particular attention.

Thanks to the modern technical aides offered by the aquarium trade today, it is no longer a problem to keep a naturally decorated aquarium operating without unreasonable effort. The water-values recorded in the natural biotopes can even then be accomplished when the available tap-water has completely adverse properties. Desalination devices are able to decrease the hardness of water to the desired values and different filter media change the pH from neutral or alkaline to acidic or vice versa.

Finally, if a sufficient and varying diet is supplied to the fish and the aforesaid recommendations have duly been taken into consideration, there is no reason why large American Cichlids should not stay healthy over many years and reproduce. In an aquarium that is set up according to the conditions found in the natural habitats there is also no point in removing eggs, larvae, or fry from the parental fish and rear them artificially. A biologically sensible consolidation of fishes and a purposeful decoration of the tank alone will already ensure that a reasonable number of offspring will grow up.

The Genus Acarichthys
EIGENMANN, 1912

This genus is monotypical, which means it contains only a single species. The separate position of this species in the systematic arrangement of Cichlids is based on the fact that it shares quite a number of features that are found in the Geophagines, e.g. in representatives of the genera *Geophagus* or *Biotodoma*. These include the shape of the body, particulars of the colour-pattern, and small scales. On the other hand, *Acarichthys heckelii* lacks the meaty appendix on the upper end of the first gill-arch that is typical of all Geophagines. Further features that led to the separation of this form from all other South American genera of Cichlids are the low maximum number of thirteen spines in the caudal fin and the peculiar structure of the first soft ray in the pelvic fins.

Particularly close affinities exist with the species of the genus *Guianacara* which are distributed in the northeast of South America, i.e. in the Guyanas, Venezuela and Brazil. Before these were eventually consolidated in a separate genus by KULLANDER in 1989, they had been assigned to the genus *Acarichthys* too.

◗ *Acarichthys heckelii*
(MÜLLER & TROSCHEL, 1849)

Localities of *Acarichthys heckelii*

Originally this species was allocated to the genus *Acara* with *Acara subocularis* COPE, 1878, being a junior synonym. The scientific literature occasionally listed it under the name *Geophagus thayeri* STEINDACHNER, 1875.

This Cichlid may reach approximately 20 cm in length. Although it does not display very striking colours, it ranks among the most splendid South American Cichlids. Eight or nine horizontal rows of glossy,

Acarichthys heckelii ♂

greenish golden spots are arranged on the light grey background colour of the flanks complemented by a larger black spot in the centre of the sides. Adult specimens often show a bright orange brownish zone between this blotch and the gill-cover that extends from the back down to the chest. The same colour is found on the cheeks and in the pelvic and anal fins. The upper anterior quarter of the eye is blood-red and so are the conspicuous threadlike appendices of the first five or six rays in the soft part of the dorsal. Up to a length of four centimetres, juveniles show a very characteristic colour feature in the anterior portion of the dorsal fin, i.e. the first two membranes are deep black followed by two bright red ones. Sexes are difficult to distinguish, but old males develop more expansive fins.

Specific traits

Important recognition characteristics include the round lateral blotch situated a half to one scale below the lateral line, the filamentous prolongations of the anterior rays of the soft dorsal fin, and the large, bright orange brown coloured zone immediately behind the gill-cover.

Similar species

A resembling extension of the rays in the soft dorsal fin is found in *Satanoperca daemon* (HECKEL, 1840), but this species is considerably more slender and shows a pitch black spot on the caudal peduncle. As far as colours are concerned, there are also similarities to representatives of the genus *Geophagus,* which however lack those produced rays in the soft caudal fin.

A similar orange brown colour-zone on the shoulders is present in *Biotodoma cupido* (HECKEL, 1840). In this species, the lateral blotch is not found in the centre of the flank, but on the posterior half of the body, closer to the caudal fin.

There is also a certain resemblance of the species of the genus *Guianacara* which should only be a cause for misidentification in the case of juveniles because the young fish of some species also have conspicuous markings in the anterior part of the dorsal find similar to *Acarichthys heckelii*. Fully grown species of *Guianacara* then have completely different body- and fin-shapes.

One of the authors had an opportunity to examine the

natural habitats

of this Cichlid in the Rio Negro region where it occurs in calm side rivers and backwaters with ample hiding-places in the form of submerged logs and branches. In some places, it was the commonest Cichlid. It mostly occurred together with *Mesonauta insignis* and *Biotodoma wavrini* in this area. In the Rio Negro drainage, the fish lives in extreme black water with a total and carbonate hardness below 1 °dH and a pH usually around 5. It appears that they inhabit an expansive area and it is therefore not unlikely that they also occur in waters with different water chemistries. Besides Rio Negro, the literature mentions localities such as Rio Juruá, Lago Hianuari near Obidos, Rio Tapajós, Rio Tefé, Rio Xingu, and Rio Trombetas in Brazil, Essequibo and Rupununi Rivers in British Guyana, as well as the river Pebas.

Juvenile *A. heckelii* with their typical pattern

The natural habitat of *Acarichthys heckelii* in the drainage of the Rio Negro (Brazil). The fish prefers a slow current and many hiding places.

The successful long-term

care

of this species requires a larger tank of about one and a half metres in length. To be comfortable, the fish need ample hiding-places that should be created with bog-oak and rock flakes. Planting the aquarium is possible as the fish do not damage the vegetation, but the plants should never constrain the available swimming space. Too bright lighting causes the fish to be shy and show pale colours.

A. heckelii has a very peaceful and calm disposition and should therefore be never kept together with more robust and aggressive Cichlids. Due to their size they require nutritious food in adequate quantities. Among others, peeled shrimp and crabs are suitable.

Its
breeding

in captivity has only been successful in a few instances. BARAN (1981) bred this species in ordinary tap-water with a total hardness of 16-18 °dH. These water values were certainly not ideal though. Taking into consideration the conditions in the natural biotopes, one should rather go for water poor in minerals with a pH in the acidic range.

As *A. heckelii* is a cavebrooder which attaches its eggs to the roof of a cave, the decoration of the tank requires particular attention. Suitably large flower pots with a relatively small opening appear to find ready acceptance. Both male and female care for and protect their eggs and fry by forming a parental family-structure.

The Genus Acaronia
MYERS, 1940

This genus used to be considered a mono-typical genus for a long time, but today incorporates the two species *A. nassa* and *A. vultuosa*. While the former is widely distributed in the Amazon basin and the Guyanas, the second only occurs in Colombia, Venezuela and Brazil. The genus was suggested by MYERS to replace *Acaropsis* which was originally introduced by STEINDACHNER for his species *A. nassa*. It turned out that this name was preoccupied.

The species of *Acaronia* closely resemble those of the genus *Chaetobranchus* as far as the body shape is concerned. They are characterized by a deep, protrudable mouth and very large scales.

Acaronia vultuosa: a specimen from Rio Caura shortly after capture

◆ *Acaronia vultuosa*
KULLANDER, 1989

Localities of *Acaronia vultuosa*

This fairly inconspicuous Cichlid can grow up to a length of about 18 cm. Its background colour is grey to greyish brown that is in contrast with the dark margins of the scales.

Depending on the prevailing mood, the flanks may be marked with a broad lateral stripe and six dark bands between the edge of the gill-cover and the caudal fin, the latter decreasing in width posteriorly.

The blood-red colour of the iris is conspicuous. The head region is marked with the dark spots typical of this species, but their intensity changes with the mood. The most important traits to distinguish live fish in the aquarium from *A. nassa* include the large lateral blotch that extends up to the lateral line and the lacking of a spotted pattern in the caudal fin.

Natural habitats

The range of *Acaronia vultuosa* is restricted to the drainage of the Orinoco in Venezuela and Colombia and the upper Rio Negro in

An adult specimen of *Acaronia vultusa*

Brazil. There, the fish inhabit a variety of water-bodies that are usually characterized by very soft and acidic water and a slow flow. On one collecting site on the Rio Caura we recorded the following values: water-temperature 30 °C, conductivity ‹10 µS/cm, pH 5.3.

Care

This Cichlid is calm, sometimes even shy and shows little resistance when kept together with other Cichlids. As this is a predatory fish which in nature not only feeds on shrimp and insect larvae, but also on small fish, adult specimens require a rich diet accordingly.

Breeding

Breeding attempts should only be undertaken in very soft, acidic water. If exposed to water values other than this, there is a risk that the embryos do not develop normally. *Acaronia vultuosa* is a typical openbrooder where male and female care for and protect the offspring in a parental family-structure.

The Genus Aequidens
EIGENMANN & BRAY, 1894

The generic name was suggested by the ichthyologists EIGENMANN & BRAY in 1894 to replace the name *Acara* that had been in use up to this point of time. This step had become necessary after it had turned out that *Acara* in fact was a synonym of *Astronotus*. With more than 25 species, *Aequidens* used to be the second largest genus of South American Cichlids. In his 1983 revision, KULLANDER eventually pointed out that this unit would not be monophyletic, but merely a collective pool to store Cichlids of very different clades.

In order to demonstrate the real cladistic relationships he divided the genus into a number of species-groups, the majority of which have meanwhile been granted the status of separate genera. The recently described genera containing species of the former composite genus *Aequidens* include *Bujurquina* KULLANDER, 1986, *Laetacara* KULLANDER, 1986, *Tahuantinsuyoa* KULLANDER, 1986, *Krobia* KULLANDER & NIJSSEN, 1989, *Cleithracara* KULLANDER & NIJSSEN, 1989, and *Guianacara* KULLANDER & NIJSSEN, 1989.

At present, the genus *Aequidens* contains a dozen or so recognized species which are added by a number of forms that have not yet been dealt with taxonomically. Some of these are presently kept in captivity. As far as the habitus is concerned, some species of *Aequidens* closely resemble representatives of the genus *Cichlasoma*. The latter partly differ in having more than three spines in the anal fin and that the bases of their anal and dorsal fins are covered with scales. In the species of *Aequidens*, the bases of the fins are nude.

The distribution of the genus *Aequidens* ranges from the Orinoco drainage in the north to the Rio Paraguay river system in the south.

Despite the transfer of numerous species formerly assigned to *Aequidens* to new genera, there is still a number of Cichlids which can neither be properly assigned to *Aequidens* as defined today, nor would they fit into any other genus. These Cichlids include, for example, the members of the *"Aequidens" pulcher* species-group. In the following these species are referred to as *"Aequidens"*, i.e. with the generic name in inverted commas.

◗ *"Aequidens" biseriatus*
(REGAN, 1913)

Localities of *"Aequidens" biseriatus*

When first described, this Cichlid was assigned to the genus *Cichlasoma*. It was later transferred to the composite unit *Aequidens*. As it is not closely related to the other representatives of the genus *Aequidens*, this arrangement must be regarded as being of temporary nature until a new organization becomes effective. Certain characteristics indicate that there is a closer relationship with the *"Aequidens" pulcher* species-group.

Males of this species may grow to a total length of more than 15 cm. Female specimens are considerably smaller. The fish have a light background colour which is in strong contrast with the black rear edges of the individual scales. Depending on the present mood, there may be six broad dark bands or a lateral stripe between the edge of the gill-cover and the base of the caudal fin. Immediately below the lateral line there is a clearly defined lateral blotch on the upper half of the flank. The area around the head is bright yellow with the cheeks being reticulated with shiny light green lines. A subocular stripe is present and distinct. The posterior portion of the gill-cover shows another black banded pattern. The posterior section of the dorsal fin as well as the upper edge of the caudal fin carry conspicuous orange red margins.

"Aequidens" biseriatus

Identifying the sexes is rather difficult in the case of this Cichlid as there is no distinct sexual dimorphism. Useful hints are only provided by the dorsal and anal fins that are far more produced in adult males than in females.

Natural habitats

The distribution of *"Aequidens"* biseriatus is restricted to the northwest of Colombia. The type locality lies in the Rio Condoto, a tributary to the upper Rio San Juán. Besides, this species was collected in the Rio Calima and in the drainage of the Rio Atrato.

Various authors recorded water values at collecting sites in the vicinity of the town of Istmina. They demonstrate that the fish lives in very soft water with a maximum total hardness of 1°dH and a pH around neutral.

Care

Provided with proper conditions, *"Aequidens"* biseriatus is a robust and adaptable Cichlid. The care of this rarely kept fish only requires a comparatively small tank of

Table 5:

Site:	Small stream in the drainage of the Rio San Juan a few kilometres southeast of the town of Istmina (Choco Province, Colombia) (according to HANSEN)
Clarity:	clear water
Colour:	
pH:	7.0
Total hardness:	< 1°dH
Carbonate hardness:	< 1°dH
Conductivity:	
Depth:	< 1 m
Current:	slow
Water temperature:	24.2°C
Air temperature:	29.7°C

one metre in length. Its bottom should be covered with several centimetres of fine sand. To feel home, the often somewhat shy fish need ample cover and numerous hiding-places. This necessitates that the back portion of the tank is densely planted with sturdy specimens of *Echinodorus* and *Anubias*. Pieces of bog-oak should furthermore be arranged to form niches and caves into which the fish can retreat when they feel like it.

As the relatively small mouth of this Cichlid indicates, it feeds on small invertebrates. Its diet should therefore consist of pondfood such as mosquito and other insect larvae complemented with small crustaceans of the genera *Daphnia* and *Cyclops*. Flake-food may additionally be offered.

Breeding

Breeding this species in captivity has repeatedly been successful. Like many other representatives of the tribe Cichlasomini *"Aequidens"* biseriatus is a typical openbrooder which preferably spawns on rocks or roots. Digging at the subsequent spawning site is apparently part of the courtship behaviour in this species.

During periods of parental care, the fish form a parental family-structure with no obvious allocation of certain tasks to the male and the female. After spawning, the female and also the male care for the embryos and the fry which are initially kept in a small pit. Even after the larval stage has been completed, both male and female guide and protect the school of young fish.

Rearing the juveniles is easy as they are large enough to feed on newly hatched larvae of Brine shrimp when they begin to feed on their own. Attempts to breed this species should take into consideration the conditions in the natural habitats. It is only soft and acidic water that forms a basis for an optimal development of the embryos.

▶ *Aequidens metae*
EIGENMANN, 1922

Localities of *Aequidens metae*

This fish is fairly well known in the aquarium hobby and was first imported alive into Europe by ourselves in 1982. It can reach a length of some 20 cm.

The body has beautiful colours of brownish yellow to yellowish with the colouration spreading onto the fins. Blueish green reflections overlay the lateral parts. Shiny lines of a similar colour are also present on the cheeks. The most significant black markings include a large blotch on the upper half of the caudal peduncle, and a large lateral spot below the lateral line that may extend up to the base of the dorsal fin and that is bordered to the left and to the right by vertical yellow margins. Furthermore, there is a vertical streak on the cheek

Aequidens metae

near the edge of the pre-gill cover. Occasionally, there may be indications of a dark stripe and several bands on the lateral parts of the body. The most significant

specific trait

that allows an identification of this Cichlid is the vertically arranged streak on the cheek extending from the hind edge of the eye to the lower angle of the pre-gill cover (comp. rear cover photograph).

Similar species

that may easily be confused with *Aequidens metae* first and foremost include *A. diadema* (HECKEL, 1840) which is a common species for the aquarium. The latter species lacks the streak on the cheek, but has a small cheek spot instead. It also lacks the attractive blueish green reflections. The

natural habitats

lie in the drainage of the Rio Meta whose lower section forms the border between the countries of Colombia and Venezuela. There is a possibility that the distribution also includes parts of the central Orinoco in Venezuela. The specimens which formed the basis of the original description originated from near Barrigón (Caño Carniceria, Cumaral). We found and observed *Aequidens metae* east of the town of Villavicencio in Colombia where it lives in various water-bodies that belong to the drainage of the upper Rio Meta. These were small, clear, partly slightly brownish tinged creeks with a bottom of sand or gravel. Aquatic plants were rare, but there was ample vegetation along the banks.

Analyses of the water revealed extremely low degrees of hardness and very acidic pH-values around 5. Syntopic fishes in these habitats included the Cichlids *Mesonauta insignis*, *Apistogramma macmasteri*, and *Apistogramma viejita*, as well as numerous Characins. The

care

of this Cichlid necessitates an aquarium of at least one metre in length. Its decoration should consist of fine gravel and a number of larger rocks and pieces of bog-oak that are arranged to create hiding-places. As this fish does not damage plants, their use is optional.

Aequidens metae is a placid and peaceful species which can also be kept together with fishes of smaller size. Even intraspecific quarrels are rare in a tank of appropriate size. It is advisable to regularly feed adult specimens with shrimp meat to keep them in proper condition.

It is unfortunate that their

breeding

does not seem to be an easy task. There are no chances of success if hard to moderately hard water is used. Instead, it has to be soft water that emulates the conditions found in the natural habitats. This is an openbrooder which spawns on rocks and roots. The eggs, larvae, and young fish are cared for and protected by both the male and female. Initial feedings of the offspring should consist of baby Brine shrimp and crushed flake-food.

Table 6:

Site:	Stream on the road from Villavicencio to Restrepo (Rio Meta drainage, Colombia)
Clarity:	clear
Colour:	slightly brownish
pH:	4.8
Total hardness:	$< 1°dH$
Carbonate hardness:	$< 1°dH$
Conductivity:	38 µS at 26.3 °C
Depth:	up to 0.4 m
Current:	slow
Water temperature:	26.3 °C
Air temperature:	27.0 °C
Time:	15.30 h

▶ *Aequidens pallidus*
(HECKEL, 1840)

Localities of *Aequidens pallidus*

As the distribution of this Cichlid includes water-bodies that are intensely exploited by the commercial ornamental fish trade, this fish is a regular item in shipments from Brazil. It mostly comes as a sidecatch and is offered under a variety of inappropriate names. The scientific and aquaristic literature also dealt with this species under the name *Aequidens duopunctata* HASEMAN, 1911, and at one stage *Aequidens pallidus* was considered a synonym of *Aequidens tetramerus*.

The upper half of the body has a whitish grey or brownish yellow ground colouration. The black rear margins of the scales are particularly obvious on the back. The posterior edge of the eye and the caudal peduncle are linked with a black stripe whose intensity varies with the disposition. The pitch black caudal spot on the upper half of the caudal peduncle is clearly separated from the lateral band by an orange coloured zone. A large lateral blotch lies on the posterior half of the body and extends from the lateral stripe up to above the lateral line and may even reach the base of the dorsal fin. It is accompanied both anteriorly and posteriorly by a yellow or orange coloured spot that often runs out in narrow whitish vertical lines. The iris is usually deep red.

Specific colour traits include a reticulated pattern of shiny green, worm-like lines on a reddish or brownish red background in the cheek region. There is a small, black cheek spot of oblong shape slightly behind and below the eye. Its presence or absence is a matter of excitement though. In juveniles the caudal fin is unpatterned whereas it shows a spotted pattern in adult specimens.

There are no distinct external features that separate the sexes. Adult male fish often have larger fins, and they mostly have a wider area of the lower head region covered with the reticulated pattern than is the case in females. Since adult males grow to sizes exceeding 20 cm, the fish is of some significance for food purposes where it occurs naturally.

A closely related species is *Aequidens tubicen* KULLANDER, 1991, from Rio Trombetas whose lateral stripe is disrupted and forms a number of spots. *Aequidens diadema* (HECKEL, 1840) differs with regard to the position of the lateral blotch that is not situated on the posterior half of the body but more in its centre.

Natural habitats

All collecting sites of *Aequidens pallidus* known to date lie in Brazil in the drainages of the central and upper Rio Negro and in the river systems of the Rio Preto da Eva, Rio Puraquequara, and Rio Uatumá. It is therefore quite certain that this Cichlid exclusively inhabits black water rivers. During the years 1981, 1986, and 1987, we repeatedly had opportunity to observe and catch this species in the vicinity of the Anavilhanas Islands in the lower Rio Negro. There, the water is transparent, but of a dark brown colour. It is extremely poor in

Aequidens pallidus

minerals and acidic. Adult specimens often reside near the banks where submerged logs and branches provide places to hide. Juveniles up to a length of about four centimetres were frequently discovered in the thick layer of rotting leaves that covers the bottom of those waters near the banks.

A swampy pond on one of the Anavilhanas Islands was particularly rich in fish, and *Aequidens pallidus* lived here in company of

Table 7:

Site:	Arquipélago das Anavilhanas: swampy pond on an island near the left bank of the Rio Negro
Clarity:	clear
Colour:	dark brown, tea-coloured
pH:	4.3
Total hardness:	< 1°dH
Carbonate hardn.:	< 1°dH
Conductivity:	10 μS/cm
Depth:	max. 1 m
Current:	none
Water temperature:	28°C
Air temperature:	26°C
Date:	26.3.1986
Time:	9.00 h

Acarichthys heckelii, Acaronia nassa, Satanoperca jurupari, Apistogramma pertensis, Apistogramma gephyra, and *Taeniacara candidi.*

Care

As *Aequidens pallidus* is a very placid fish which offers little resistance to more aggressive and lively species, it can be kept in the company of moderately sized Cichlids (e.g. species of *Biotodoma)* or even with Dwarf-cichlids *(Laetacara* sp.). Another positive attribute is that it does not damage plants. Due to its adult size it requires a very spacious tank with a length that clearly exceeds one metre.

Breeding

Although this species is imported on a regular basis, there are no reports on any breeding success. This probably results from the fact that reproduction can only be successful if the water values in captivity correspond with those found in the natural biotopes. This means extremely soft, acidic water.

▶ Aequidens patricki
KULLANDER, 1984

This Cichlid was dealt with in the ichthyological literature long before its formal
scientific description in 1984, but until then
it had always been mistaken for *Aequidens
tetramerus*.

Aequidens patricki certainly belongs to
the most striking representatives of this
genus. The basic colour of this fish is
brownish yellow with pale green reflections
in grown-up specimens. The most conspicuous feature as far as colours are concerned is the brownish orange to reddish
orange cheek region that is covered with
large, shiny, green blotches. Below the eye
there is a triangular black spot on the edge
of the pre-operculum. Further black markings are such as a spot on the caudal
peduncle and a lateral blotch that is usually
bordered by two narrow light bands. This
blotch may cover the upper half of the body
and reach up to the base of the dorsal fin. It
is not uncommon that it also extends down
onto the ventral side thus forming a broad
band. This is complemented by a lateral
stripe and a number of diffuse bands whose
intensities depend on a certain mood. All
unpairy fins show distinct spotted patterns.

The total length of this Cichlid may
exceed 15 centimetres. In contrast to other
species, not only the males but also the
females develop produced dorsal and anal
fins. Despite this fact the sexes are easy to
distinguish as the females display brighter
shades of red in their fins.

Natural habitats

Aequidens patricki has a fairly limited range
and is known only from a few western
tributaries to the Rio Ucayali in Peru. To
date the species has exclusively been found
in the drainage of the upper Rio Aguaytia
near the town of the same name and in the

Localities of *Aequidens patricki*

region of the Rio Pachitea in the vicinity of
the villages Puerto Bermúdez and Pan-
guana.

U. MINDE was the first to import live
specimens into Europe in 1987. He also
managed to breed this species first and
reported in 1988 that he found the fish living in a stream carrying clear, brownish coloured water. The flow was moderate and
the depth measured about half a metre. He
recorded a water-temperature of 24 °C, a
pH of 5.8, and a conductivity of 30-40 µS/
cm.

Care

Aequidens patricki is a robust, adaptable
Cichlid which does not pose any particular
demands on the chemical properties of the
water. It is best to keep this fish in a community tank as individually kept pairs are
often very shy. Suitable company fishes are
found among species of *Bujurquina*, *Cichlasoma*, and *Geophagus* which are of similar
size.

When keeping this Cichlid, it must be
borne in mind that it is a predatory fish
which considers clearly smaller fishes, such

as Tetras, an appreciated addition to its diet. It should therefore receive nutritious food including, for example, meat of fish and shrimp.

The aquarium should be decorated with sand or fine gravel, rocks, pieces of bog-oak, and strong plants such as larger specimens of *Echinodorus* or *Anubias*. The setup should also provide a number of hiding-places to which the fish may retreat when required.

Breeding

This Cichlid has successfully been bred in many instances that indicates that this is not particularly difficult and even works in moderately hard water. Problems are only encountered with the pronounced intraspecific aggression in this species. Finding a compatible pair among some adults can become a difficult task as fights among these powerful fish soon result in serious damage or even death. Therefore it is a good idea to acquire a number of juveniles and wait until a pair bonds within this group and starts courtship.

Aequidens patricki is a typical open-brooder which cares for its offspring in a parental family. Spawning takes place on a rock, a piece of wood, or the leaf of a sturdy aquatic plant. In an attempt to camouflage the eggs, the parents often spit sand over them. The freshly hatched larvae are then deposited in a pit.

At a water-temperature of about 27 °C it takes approximately three days from spawning to the hatching of the larvae, and about a week until the young fish eventually swim freely. At this stage they are already of a fair size and thus are able to feed on freshly hatched nauplii of the Brine shrimp. Rearing the initially fast growing juveniles is therefore not a problem. It comes a bit as a surprise that at a later stage and despite adequate feeding their growth rate decreases significantly. Maturity is reached at a length of approximately ten centimetres.

Aequidens patricki

▶ *Aequidens* sp.
(Jenaro Herrera)

This South American Cichlid was discovered in Peru by KULLANDER as recently as in 1981. Since no scientific description was published before the finalization of this manuscript, there is no name available to properly identify this species. We caught this fish in 1983 and could import a few live specimens into Germany for the first time. The maximum size of this species appears to be about 15 centimetres.

The most eminent components in the black pattern include a lateral stripe which runs from the posterior margin of the eye to the caudal peduncle and a large, more or less rectangular blotch on the flanks which stretches from between the ninth to the twelfth spine of the dorsal fin to about the border of the lateral stripe. The upper half of the caudal peduncle carries a roundish

Localities of *Aequidens* sp. (Jenaro Herrera)

black spot. A vertically arranged black streak begins at the posterior edge of the eye and runs down towards the gular region without actually reaching it. The cheeks and gill-covers show a beautiful greenish gleam.

Aequidens sp. (Jenaro Herrera)

38

The sides of the body below the lateral stripe also have a metallic green sheen while the upper half is brownish. All unpairy fins are pale reddish. Since there is no sexual dimorphism present, the sexes can hardly be differentiated.

Specific traits

of this Cichlid that are useful for the identification of live fish in an aquarium are mainly found in the colour-pattern. Typical features are the lateral stripe, the conspicuous spot on the flanks that extends onto the dorsal fin, and the metallic green reflection on the lower body in combination with the reddish fins.

Similar species

mainly encompass *Aequidens pallidus* (HECKEL, 1840), *A. diadema* (HECKEL, 1840), and *A. metae* EIGENMANN, 1922. These Cichlids have very similar body shapes and also show a conspicuous lateral blotch. However, adult specimens of these species either lack the black lateral stripe or the distinct green sheen on the lower body parts or the red colouration of the fins.

Table 8:

Site:	Black water creek 13.5 km east of Jenaro Herrera on the road to Colonia Agamos (Depto. Loreto, Peru)
Clarity:	clear
Colour:	dark brown
pH:	5.4
Total hardness:	< 1°dH
Carbonate hardness:	< 1°dH
Conductivity:	3 µS at 27.5 °C
Nitrite:	< 1 mg/l
Depth:	up to 1 m
Current:	slow
Water temperature:	27.5 °C
Air temperature:	32.0 °C
Date:	28. 6. 1984
Time:	17.00 h

The

natural habitat

of this *Aequidens*-species lies in the drainage of the lower Ucayali in northeastern Peru. The only localities known to date are in the vicinity of the village of Jenaro Herrera. We caught the fish in a small black water creek whose bottom was thickly covered with rotting leaves. Plenty of submerged branches and twigs provided the Cichlids with ample hiding-places. Juveniles were particularly numerous in this biotope. Since the creek was also home to large numbers of small freshwater shrimp it is to be presumed that these play an important role in the diet of the Cichlids. No aquatic plants were found. Other fishes inhabiting this water-body were the Cichlids *Laetacara flavilabris*, *Apistogramma nijsseni*, and *Crenicichla* sp. alongside whom we found the Characins *Crenuchus spilurus*, *Pyrrhulina brevis*, *P. laeta*, and *Curimata* sp. The

care

of this species of *Aequidens* should take place in a tank of at least one metre in length. The decoration should consist of washed fine gravel or sand, larger rocks, and pieces of bog-oak, the latter being arranged to create hiding-places. Since this Cichlid tolerates plants, those may as well be used for setting up the aquarium.

While the fish can well be kept at moderate levels of hardness, their successful

breeding

apparently requires softer water with an acidic pH. Up to the finalization of this manuscript, it could not be managed to breed this species in captivity. It is therefore still unknown whether it belongs to the openbrooders or the larvophile mouthbrooders. The genus *Aequidens* contains both types of reproduction and parental care.

▶ "Aequidens" sp.
(Silverseam-cichlid)

This Cichlid was used to be taken for *"Aequidens" rivulatus* (GÜNTHER, 1859) during the seventies. Subsequently it was discovered that this fish from northwestern Peru represents a new undescribed species.

This fish may attain a length of approximately 25 cm. The most eminent attributes in its colour-pattern are a small black spot in the centre of the caudal peduncle, a black lateral blotch that lies immediately below the upper lateral line and is bordered with a silvery vertical line on either side, a broad white margin on the posterior edge of the caudal fin, and a narrow white seam on the dorsal fin. The scales on the flanks have greenish golden reflections. The fins have shiny blueish green spots and dashes.

An indication of a sexual dimorphism appears only with age when the larger males develop a deeper forehead and longer fins. The

Localities of *"Aequidens"* sp.

specific traits

by which an identification of this Cichlid becomes possible consist of the whitish silvery borders of the lateral blotch and the posterior white margin of the caudal fin. Furthermore, it has scales with a greenish reflection, a pale centre, and black edges.

"Aequidens" sp. ♂

Similar species

are found in the *"Aequidens" pulcher* species-group which, according to KUL-LANDER (1983), also includes *"Aequidens" rivulatus*. Confusion appears to be most likely to happen in the case of another undescribed species which aquarists usually refer to as the "Goldseam-cichlid". This species has more or less the same body shape and an almost identical colour-pattern. The difference lies in the caudal fin that has an orange margin, and the dark

This floating carpet of Water hyacinths, *Eichhornia azurea*, provides cover from predators and thus is home to many Cichlids.

edges of the scales are replaced by a dark spot so that there is no reticulated pattern on the flanks, but spots that appear to be arranged in horizontal rows.

Natural habitats

The distribution of the Silverseam-cichlid extends over the water-bodies along the Pacific coast of northwestern Peru. Relevant literature mentions localities south of the Rio Tumbes in the vicinity of the villages of Piura, Pacasmayo, and the wider surroundings of Lima (Huacho). LÜLING (1973) caught this fish in a lagoon with brackish water where both total and carbonate hardness measured 6.7 °dH. Conductivity and pH were established by him to be 4280 micro-Siemens and 7.7 respectively.

These values demonstrate that this Cichlid is highly adaptable and the water quality is of secondary importance as far as its

care

is concerned. Problems may be encountered when it comes to the composition of compatible pairs as the individual specimens are highly aggressive towards each other and require ample space. During periods of parental care, this Cichlid may also become extraordinarily vicious towards other Cichlids in the same tank. Its long-term husbandry therefore necessitates an aquarium of at least one and a half metres in length. It should only be kept together with equally valiant species.

The tank should be decorated with pieces of bog-oak and rocks that create sanctuaries and simplify the definition of territorial borders. This fish requires rich food in the form of cut fish- and shrimp-meat.

Breeding

is easy as compatible pairs willingly spawn under a variety of keeping conditions. This Cichlid is an openbrooder which rears its offspring in a parental family-structure.

▶ *Aequidens* sp. aff. *tetramerus*
(Ecuador)

The Swedish ichthyologist KULLANDER (1986), to whom also fishes from the drainage of the upper Rio Napo were available, considers this Cichlid from Ecuador to be *Aequidens tetramerus* (HECKEL, 1840). Said species was originally described from the Rio Branco in Brazil. As it is presently understood, it is fairly variable and has an unusually wide range. Despite the assignment by KULLANDER, there are indications that this Cichlid from Ecuador is in fact a separate species. KULLANDER obviously failed to take into consideration the colours in life as he had only conserved specimens available.

It is a peculiarly coloured *Aequidens* which has been imported into Germany on a number of occasions. It also has been bred successfully (STAWIKOWSKI & WERNER 1988). We brought home a few live specimens from a trip to Ecuador in 1990 and deposited

Localities of *Aequidens* sp. aff. *tetramerus*

conserved material in the Museum für Naturkunde in Berlin (ZMB 32310).

Adult specimens of this species display a characteristic colour-pattern that makes it difficult to confuse them with any other South American Cichlid. The most signifi-

Aequidens sp. aff. *tetramerus* from Ecuador

cant colour trait is an intense wine- or car-mine-red zone that extends over the lower head and chest region even reaching the anterior portion of the belly in some indi-viduals. This red colour becomes apparent when the fish has grown to about five cen-timetres. The rest of the body is more or less blueish grey. The iris is golden with the upper half also being red. There is a large, pitch black blotch on the upper half of the caudal peduncle that often stretches onto the lower half and then rather looks like a broad band. This black marking is bordered with a narrow golden rim. Influenced by certain moods, the flanks may show a nar-row dark stripe besides some bands. Such a stripe would then run from the posterior edge of the eye to the caudal blotch. It may be complemented by a large lateral blotch that often expands vertically to form a broad band.

The lower halves of the head and the body of adult specimens have a beautiful pale green iridescence. The fins are blueish grey, but can also have a reddish hue. The posterior portions of the anal and caudal fins are patterned with short, narrow, green, shiny dashes.

The above description of the colour applies to specimens of this species caught in the drainage of the Rio Coca in the vicinity of the town of the same name. Spe-cimens from Rio Aguarico lack the green sheen on the flanks as well as the distinct pattern in the anal and caudal fins (STAWI-KOWSKI & WERNER 1988).

The maximum total length of this species is about fifteen centimetres. As there is no well-defined sexual dimorphism, the usually more pointed and more heavily marked fins of the males are the only traits that provide a hint about the sex.

Natural habitats

The distribution of this Cichlid appears to be confined to the drainage of the upper Rio Napo (Rio Coca, Rio Aguarico). In February 1990, we caught this fish in the Rio Yanayacu, a tributary to the Rio Coca, besides the road to Lago Agrio, some 20 kilometres north of the town of Coca. This small river had a considerable current and a width of about five to fifteen metres at this site. Its water was crystal clear, soft, and slightly acidic. Even in its deeper sections the maximum depth hardly reached one metre. The river bed therefore had a num-ber of sandbanks that were as densely over-grown with *Myriophyllum matogrossense* as many sections of the banks. These plants provided cover for numerous fish. Besides *Aequidens* sp. aff. *tetramerus* we also caught a species of *Mesonauta* and *Crenicichla* at this site.

Care

This Cichlid does not confront its keeper with specific demands, but can be kept like other species of *Aequidens*. It belongs to the typical openbrooders which spawn on a hard surface and care for and protect their eggs, larvae, and young fish in a parental family-structure. Breeding was also suc-cessful in moderately hard water. The fry can initially be fed with freshly hatched Brine shrimp.

Table 9:

Site:	Rio Yanacu on the road to Lago Agrio, about 20 km north of the Town of Coca.	
Clarity:	very clear	
Colour:	yellowish	
pH:	6.8	
Total hardness:	$< 1\,°dH$	
Carbonate hardness:	$2 - 3\,°dH$	
Conductivity:	$50\,\mu S/cm$	
Depth:	$< 1\,m$	
Current:	swift	
Water temperature:	$25\,°C$	
Air temperature:	$29\,°C$	
Date:	8.2.1990	
Time:	11.00 h	

◗ *Aequidens uniocellatus*
(CASTELNAU, 1885)

Localities of *Aequidens uniocellatus*

was originally included in the genus *Chromis*. A junior synonym is *Acaronia trimaculata* ALLEN, 1942. Various authors considered this species to be a synonym of *Aequidens tetramerus* (HECKEL, 1840). The fish reaches a maximum length of nearly 20 cm.

Significant traits of the colour-pattern of this Cichlid are a large black lateral blotch and a large caudal blotch. The latter is situated on the upper half of the caudal peduncle and has a frame of silvery whitish scales. Individual scales of this kind are often also present on the lower parts of the caudal peduncle. The iris is red to reddish orange. Below the eye there is a small black cheek spot. Chest and gular regions are yellowish to bright yellow. Certain moods produce a dark lateral stripe on the flanks or a number of dark bands. This fish is iso-

Aequidens uniocellatus

morphic which means that live males and females are not distinguishable with certainty.

Specific traits

that allow an identification of *Aequidens uniocellatus* and a separation from similar species include the large caudal blotch being bordered silvery, the silvery scales on the caudal peduncle and the yellow colour of the head and chest regions.

The

natural habitats

of *A. uniocellatus* lie in Peru, in particular in the drainage of the Ucayali River. We could find this species in 1983, in the region of the central Ucayali in the vicinity of Pucallpa, but also in a small tributary of the Rio Tambo. One collecting site was a small creek that was thickly vegetated by *Echinodorus*, water-lilies, and other plants and flowed into the small clear water stream. Another locality was an isolated pool filled with dead leaves and branches and belonged to a small white water river. Analyses of the water revealed that its pH ranged at slightly alkaline levels with maximum recordings of 13.5 °dH total and 16.5 °dH carbonate hardness. Water-temperatures varied from 24 to 28.4 °C.

Fishes occurring at the same localities other than *A. uniocellatus* were such as *Mesonauta mirificus*, *Pterophyllum scalare*, *Cichlasoma amazonarum*, *Apistogramma cacatuoides*, and catfishes representing the genera *Farlowella*, *Hypoptopoma*, and *Bunocephalus*.

Observations made in the natural biotopes suggest that the

care

of this Cichlid is not difficult. The length of the tank should not fall short of one metre.

Its decoration should include hiding-places and visual barriers by using pieces of bog-oak, rock flakes, and some dense groups of plants to make the fish feel at ease. The chemical qualities of the water are not of particular importance. *A. uniocellatus* is relatively timid and can well be kept together with other South American Cichlids. Being an omnivore, feeding is no problem whatsoever.

Breeding

occurs without special conditioning. The fish also spawn willingly in a community tank if it is not overcrowded. They show the behaviour typical of openbrooders with both parental fish caring for and protecting the eggs, larvae, and young fish. Ground flake-food and baby Brine shrimp form the basis of initial feedings.

Table 10:

Site:	White water creek in the drainage of the Paca Cocha, or Yarina Cocha respectively, near Pucallpa (Peru).
Clarity:	very turbid
Colour:	loam-yellow
pH:	7.6
Total hardness:	13.5 °dH
Carbonate hardness:	16.0 °dH
Conductivity:	499 µS at 29 °C
Depth:	up to 1 m
Current:	slow
Water temperature:	29.0 °C
Air temperature:	28.4 °C
Date:	29. 6. 1983
Time:	16.00 h

The Genus Amphilophus
AGASSIZ, 1858

For a long time this generic name was a synonym of *Cichlasoma*. AGASSIZ had introduced it when he described *Amphilophus froebelii*, which is regarded a synonym of *Amphilophus labiatus* today. The reorganization of the heterogeneous genus *Cichlasoma* and its restriction to South American Cichlids (KULLANDER 1983) necessitated that all Central American species included herein formerly had to be assigned to other genera. The genus *Amphilophus* presently holds those species that formed the section *Astatheros* of REGAN (1906-1908). It requires a revision urgently. Due to the fact that it represents a very heterogeneous unit, various authors distinguish between several species-groups.

Localities of *Amphilophus alfari*

◆ Amphilophus alfari
(MEEK, 1907)

A synonym of this Central American Cichlid is *"Cichlasoma" lethrinus* REGAN, 1908. It was imported into Germany for this first time by WERNER in 1981 who caught specimens in the Rio Pacuare, Rio Santa Maria, and Rio Blanco (Costa Rica). The maximum total length recorded for free-ranging specimens is 17 centimetres.

This fish's body is greyish brown on the back and pale grey on the belly. The lower portion of the head is orange, while the gular and chest regions are pink. Six diffuse dark bands may be present on the flanks, complemented by a dark lateral stripe that connects the eye with the caudal peduncle. At the point where the lateral stripe crosses the third band, there is a deep black spot right below the lateral line. The caudal blotch is usually reduced to a faint indication. A number of glossy blueish green spots are found below the eye and on the

gill-cover. The dorsal fin is marked with a conspicuous signal-red margin while the ventral fins are yellowish orange. Specimens from certain localities show a bright red colouration of the lower half of the caudal fin and the posterior portions of the anal and dorsal fins. Due to the lack of a distinct sexual dimorphism, the sexes are difficult to distinguish.

Specific traits

that may identify this Cichlid include the pattern consisting of a dark lateral stripe in conjunction with several lateral bands, the orange coloured head, and the pink gular and chest region. Furthermore, this species usually has 17 to 18 spines in the dorsal fin. It is noteworthy that populations from the Atlantic slope differ considerably from those from the Pacific counterpart which made BUSSING (1966) consider them a separate subspecies. The Atlantic subspecies is illustrated here. The Pacific form differs in being more slender, having a lower number of glossy spots on the cheeks and gill-covers, and having an incomplete lateral stripe that is usually confined to the anterior part of the body.

Amphilophus alfari

A

similar species

is primarily *Amphilophus diquis* BUSSING, 1974. This species has partly forked lateral bands, it lacks the blueish green reflections, and its chest region is not pink. The number of spines in the dorsal fin is usually limited to 16.

On the Atlantic side of Central America, *Amphilophus alfari* is distributed from the east of Honduras through Nicaragua and Costa Rica to western Panama. On the Pacific slope, the

natural habitats

all lie in Costa Rica. Localities are primarily known from the drainage systems of the Rio Bebedero and Rio Tarcoles. On the Atlantic side this Cichlid is found in the drainages of the rivers Madre de Dios, Reventazón, Sarapiqui, San Carlos, and Arenal, to name just some localities. The fish inhabits soft as well as moderately hard water with a pH ranging from slightly acidic to slightly alkaline.

Therefore, as far as the

care

of this Cichlid in captivity is concerned, the chemical properties of the water are not very important. A tank measuring about one metre in length is appropriate. It should be decorated with large rocks and pieces of bog-oak in such way that sanctuaries are created and the definition of territorial borders is simplified. Hardy aquatic plants may be used. This species can be kept together with other Central American Cichlids of comparable size, e.g. species of the genera *Archocentrus* or *Thorichthys*. Fully grown fish require a nutritious diet including pieces of fish and shrimp.

Breeding

this species in captivity is not difficult. *Amphilophus alfari* is an openbrooder. Both male and female care for the eggs, larvae, and young fish and defend them when necessary. Initial feedings should consist of freshly hatched larvae of *Artemia salina* and pulverized flake-food.

47

◗ *Amphilophus citrinellus*
(GÜNTHER, 1864)

Localities of *Amphilophus citrinellus*

Originally this Cichlid was assigned to the genus *Heros*. Junior synonyms are *Heros basilaris* GILL & BRANSFORD, 1877, and *Cichlasoma granadense* MEEK, 1907. The maximum length of this fish is 30 cm.

The background colour consists of greyish brown to olive tones. A black spot is present slightly above the centre of the base of the caudal. The presence or absence of other black markings depends very much on the prevailing mood. This includes seven dark bands and a lateral stripe that is composed of a series of six black spots, the third of which being the largest. Occasionally only this spot remains visible. *Amphilophus citrinellus* belongs to the small group of Central American Cichlids that are polychromatic, which means there is a vast variety of colour morphs within the species. The colour-pattern described afore there-fore refers to the normal variety which is encountered most often and represents about 70 % of specimens in most of the wild populations examined. A small number of fish, usually not exceeding 10 % of the total population, is noteworthy for its

Amphilophus citrinellus; male of the whitish morph

bright colours. Body and fins are whitish, yellow, copper red, pink, or orange in these specimens. The remaining specimens display the colour-pattern of the normal variety that is overlain with coloured zones.

The juveniles of all morphs show the normal pattern. Fully grown specimens of *Amphilophus citrinellus* have a distinct sexual dimorphism that makes the determination of sexes easy. Male fish develop a heavily bulged forehead. According to BARLOW (1976) this development is reversible though. The growth of the bulge appears to be connected to sexual and aggressive behavioural patterns and may progress quite rapidly. The largest bulges are thus found among isolated groups of males during pair formation periods.

Specific traits

include the lateral series of six black spots of which the third is the largest one, the distinct bulged forehead in older males, and the occurrence of various colour morphs.

Similar species

Amphilophus labiatus (GÜNTHER, 1864) is more elongate in its built, has a more pointed head, shorter fins, and usually thicker lips. This species entails a bright red colour morph. Additionally, the bulged forehead of the males is less pronounced than in *A. citrinellus*. Despite the aforesaid, both species can sometimes not be told apart with certainty if there is no informa-

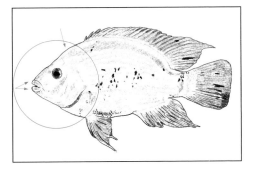

Amphilophus citrinellus

tion on localities as their characteristics overlap.

Another similar species is *A. zaliosus* BARLOW, 1976, which is considerably more slender though. It has a large, horizontally oval, black spot on the flanks.

The

natural habitats

of *A. citrinellus* lie in the two Central American states of Nicaragua and Costa Rica. Localities are known from the lakes Masaya, Jiloa, Apoyo, Managua, and Nicaragua, as well as from their tributaries Ochomogo, Sinecapa, and Rio San Juan. In Costa Rica, the fish has been recorded from the area of the Rio Frio and the Tortuguero region. In Lake Nicaragua, this species was still observed at a depth of 30 metres.

The availability of hiding-places appears to determine the distribution of this Cichlid. It is preferably found in the cobble and rocky zones of its lakes or among submerged branchwork and logs. Analyses of stomach contents revealed that the fish often feeds on plant material. Besides algae, small crustaceans, insect larvae, and snails are preyed upon. Larger specimens also feed on fish. The waters inhabited by *A. citrinellus* are usually of a neutral to distinctly alkaline quality. At a temperature of 25 °C, a conductivity of between 200 and 5500 micro-Siemens was recorded.

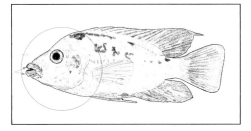

Amphilophus labiatus

Fish are often caught with a throw-seine in South and Central America.

This Cichlid is hardy and adaptable so that its

care

is easy. However, in the long run, very spacious aquaria are required whose length should not fall short of one and a half metres. The tank should be decorated with large rocks in such a way that a number of hiding-places are created. The usage of plants is limited to very robust species as the fish often feed on more delicate vegetation. It can turn out to be a problem to find a compatible pair as individual specimens may be quite aggressive towards each other. Larger fish require a rich diet that should contain meat of fish and shrimp. The intense colouration of the yellow and orange morphs largely depends on the supply of carotinoids and canthaxanthine in the food. As soon as this aspect is neglected, the colours fade. Regular offerings of adult Brine shrimp are adequate to better the situation.

Breeding

is not a problem if you have a compatible pair available. The fish usually spawn on a flat stone, more rarely they choose a cave. The larvae are subsequently deposited in pits. Both parental specimens participate in rearing the fry.

▶ *Amphilophus labiatus*
(GÜNTHER, 1864)

Localities of *Amphilophus labiatus*

was assigned to the genus *Heros* when first described in a manner that was acceptable as an original description. AGASSIZ had already dealt with this species under the name *Amphilophus froebelii* in 1859. Since his characterization, however, did not suffice the internationally valid rules of nomenclature imposed on original descriptions, this name is considered a senior synonym. With *A. labiatus* being a polymorphous species which consequently can look very different, various varieties of this species were erroneously described as separate species. Such junior synonyms include *Heros lobochilus* and *H. erythraeus* GÜNTHER, 1869, and *Cichlasoma dorsatum* MEEK, 1907. Its maximum length is almost 30 cm.

This fish has a grey to greyish green background colour. Dark patterns can appear, but largely depend on certain stages of excitement. The most important features are such as a series of six dark spots that connect the edge of the gill-cover and the caudal spot. Furthermore the flanks are marked with seven dark bands. Except the caudal and one lateral spots, all these elements may temporarily disappear.

Orange coloured morph of *Amphilophus labiatus*

Amphilophus labiatus is a polychromatic species. Besides the normal morph described afore to which about 90 % of the specimens observed and caught in the wild belong, there are several other, very distinctive varieties. These may be pink, red, or checked red and white. Often these specimens show some pitch black pigmentation in their fins, on the lips, or on the back. Unfortunately, the fantastic colours of wild-caught fish depend on certain food-items with the red colours being produced by canthaxanthine and the yellow ones by carotinoids. If these substances are not constantly included in their captive diet, the colours fade.

Experiments by WEBER, BARLOW & BRUSCH (1973) showed that regular supplies with beta-carotene preparation, red pepper, and especially adult Brine shrimp are adequate to maintain the natural colouration or even restore it. Juveniles of the colourful varieties look as drab as the normal morph, with the colour change only taking place at an age of three to twelve months. Adult males can be recognized by their enlarged fins and a deeper forehead.

The most significant

specific traits

of *A. labiatus* are the thickened, distended lips that in some specimens may have a long, triangular flap in the centre. According to BARLOW & MUNSAY (1976), the development of this extension is not only genetically fixed, but also depends on environmental conditions. It was observed that this bizarre lobe on the lips decreases after some time of keeping in captivity. It is therefore obvious that certain external stimuli that normally induce growth of the lips are absent in the aquarium.

A. labiatus is typical for producing a red colour morph which is unknown in other polychromatic species.

Similar species

are found in the genus *Amphilophus*. It is particularly difficult to differentiate between this species and *A. citrinellus* (GÜNTHER, 1864). In general, the latter species has thinner lips, a deeper body, and a less pointed head, but these traits are subject to great variability. They therefore often overlap in these species so that determination is impossible when there is no information on the geographic origin of a specimen.

The

natural habitats

of this Cichlid lie in Lake Managua and Lake Nicaragua in the Central American state of Nicaragua. At a water-temperature of 25 °C, conductivity was recorded to be 414 micro-Siemens in Lake Managua and 200 micro-Siemens in Lake Nicaragua with a pH around 8.5 in both lakes (BARLOW et al. 1976). *Amphilophus labiatus* prefers to reside above rocky bottom where it apparently also feeds on plant material. BARLOW & MUNSEY (1976) presume that they extract their prey mainly from crevices in the rocks.

For the

care

of this Cichlid a large tank of more than a metre in length is a must. It should be decorated with large rocks arranged to form several caves and territories. The usage of plants is limited by the fact that the fish often destroy more delicate vegetation. If individual specimens are kept under crowded conditions they may become very aggressive. Suitable co-inhabitants of such a tank should preferably be chosen from among other large Cichlids of the tribe Cichlasomini. *A. labiatus* is an enduring, adaptable aquarium fish.

Breeding

is not particularly difficult. *Amphilophus labiatus* follows the behavioural pattern of a typical openbrooder, but will also spawn in a cave if there is opportunity.

The offspring is reared in a father-mother-family, in which both sexes have different tasks. During the development period of the eggs and larvae, the female mainly cares for the brood while the male defends the borders of the breeding territory.

The Genus Archocentrus
GILL & BRANSFORD, 1877

For decades, this name was regarded as a synonym of the genus *Cichlasoma*. It was first used in 1877, when the ichthyologists T. GILL and J. F. BRANSFORD described *Archocentrus centrachus*. It originally served to distinguish a new subgenus within the genus *Heros*. The English ichthyologist REGAN then used this name again in 1905 to group a number of Central American Cichlids within the genus *Cichlasoma*.

Splitting and redefinition of the former catch-all genus *Cichlasoma* by the Swedish ichthyologist KULLANDER in 1983 resulted in a separation of all Central American species from their South American counterparts. It also caused the necessity to accommodate the former in other genera. In the majority of cases there were names available that so far had been synonyms of *Cichlasoma*, but had become available again by the revision of this genus. That the name *Archocentrus* would be available for use for a clade of Central American Cichlids according to its definition by REGAN (1905), was pointed out by KULLANDER already in 1983. At that stage however, he failed to formally reintroduce the name in its new function.

This step was eventually taken by ALLGAYER in 1994 on occasion of the description of a new species of *Archocentrus* from Panama. In his definition of the genus he explicitly referred to the generic traits listed by REGAN.

This genus is of some relevance for the aquarium hobby and presently holds eight species, i.e. *Archocentrus centrachus* (GILL & BRANSFORD, 1877), *A. spilurus* (GÜNTHER, 1862), *A. nigrofasciatus* (GÜNTHER, 1866), *A. spinosissimus* (VAILLANT & PELLEGRIN, 1902), *A. septemfasciatus* (REGAN, 1908), *A. cutteri* (FOWLER, 1932), *A. sajica* (BUSSING, 1974), and *A. nanoluteus* ALLGAYER, 1994. The distribution of the genus *Archocentrus* is restricted to Central American waters between Guatemala in the northwest and Panama in the southeast.

The species of *Archocentrus* are characterized by a very small adult size. In general, its males are slightly larger than the females and measure between nearly ten and fifteen centimetres. These fishes have a fairly deep body and a small mouth which cannot be protruded very far. Further generic traits are such as the enlarged front pair of teeth and the high number of eight to twelve spines in the anal fin. In almost all species there is not only a distinct sexual dimorphism, but the sexes are also differently coloured.

The genus *Archocentrus* is of importance for the aquarium hobby in so far that these small Cichlids with their very contrasting colour-patterns are almost ideal for being kept in a tank. Even under the most different keeping condition they will spawn willingly. Their hardiness and adaptability also make them highly recommendable for the beginner.

◆ *Archocentrus sajica*

(BUSSING, 1974)

Localities of *Archocentrus sajica*

This Cichlid belongs to the small number of species that were discovered and described only recently. Male specimens may grow up to a length of nearly twelve centimetres. Their female counterparts are considerably smaller and fully grown at about eight centimetres.

As there are distinct differences in the colour-pattern, males and females of this species are easily told apart. Male fish have a loam-yellow to fawn coloured back region. Their heads are bright yellowish brown to orange with the area between the gill-cover and the pre-gill cover contrasting in light blue. The rest of the body is greyish blue to greyish brown. The rays of all unpairy fins are deep wine-red while the membranes in between are iridescent pale blue. The dorsal fin has a fine red margin.

Female (bottom) and male (top) of *Archocentrus sajica*

The females of this species are less spectacular. They are mainly grey with the lower body half tending towards blueish grey. The lower head and gular region may assume a pale blue tinge. During periods of parental care the females become very dark. At these times the head, chest, and belly may become almost black.

Depending on the mood, both sexes may display seven diffuse dark bands on the flanks. The third band is usually the most prominent one. It is enhanced in the case of a breeding female to form a deep black band that extends even onto the dorsal fin. Moreover, there is a dark lateral stripe with vague borders that may run from the edge of the gill-cover up to the caudal peduncle, but usually ends at the third lateral band creating a horizontal T-shaped pattern. In addition to this, there is a diffuse black caudal spot.

Specific traits

that identify this Cichlid include the low number of six to seven spines in the anal fin. Furthermore, a well pronounced third lateral band and a dark stripe on the flanks are typical of this species.

Similar species

are especially found among the other seven representatives of this genus. *Archocentrus spinosissimus* (VAILLANT & PELLEGRIN, 1902) has a uniformly grey basic colouration, which means that there are no yellow and bright blue shades. This species also has 11 or 12 spines in the anal fin and lacks a lateral stripe. *Archocentrus spilurus* (GÜNTHER, 1862) has 8 to 10 anal fin spines. This species lacks red colours in the fins nor does it have a dark stripe on the sides of the body.

In comparison, *Archocentrus sajica* has a fairly limited distribution. Its

natural habitats

are exclusive to the Pacific slope of Costa Rica. Here, the fish has so far only been recorded from between the drainage of the Rio Esquinas and Punta Mala. Some information on the ecology of this Cichlid was published by BUSSING (1974) on occasion of the original description. *A. sajica* is exclusive to slow flowing, small water-bodies of only a few metres in width. It was found above sandy bottoms as well as above gravel. An analysis of the stomach contents revealed that the fish preferably feeds on plant material. Besides filamentous algae and small seeds, water insects and small fish were also found.

A. sajica is an adaptable species whose successful

care

must be termed easy. A tank with a length of nearly one metre provides adequate space for the fish to feel at home. Their well-being is unquestionably enhanced by the presence of cave-like shelters. Although they feed on plant material in nature, they will not damage the vegetation if explicitly delicate plants are forgone. They are relatively timid Cichlids and should therefore be only kept together with other Central American species of similar size and behavioural patterns.

Breeding

can be successful even in a community tank. Caves form the preferred spawning sites. During periods of parental care, the fish constitute a father-mother-family which means both sexes assume different duties. When the larvae have hatched, both parents engage in frantic digging activities to create pits in which the larvae can be deposited and changed around until they can swim. Using the usual diet, their rearing is easy.

▶ *Archocentrus septemfasciatus*
(REGAN, 1908)

MEEK (1914) and various later authors erroneously considered this species as identical with *Archocentrus spilurus* (GÜNTHER, 1862) from Guatemala. This misconception was among others corrected by BUSSING (1974). The fish made its first appearance in the aquarium hobby in the late seventies. Initially available only in the USA, it later also found its way to Germany. The maximum body length of the males is nearly 12 centimetres. Female specimens are much smaller and fully grown at a length of about 8 centimetres.

Localities of *Archocentrus septemfasciatus*

There is not only a sexual dimorphism in adult specimens, but also a sexual dichromatism so that distinguishing between the sexes is easy. Females are much more colourful than males. Their basic colouration consists of pale blueish grey or greyish blue shades. The upper parts of the cheeks and the gill-cover are usually yellowish to orange. The throat and chest are blue. Individual scales of the anterior half of the body show feeble coppery reflections. The flanks are marked with seven dark bands, the third of which extends onto the dorsal fin. This one and the posteriormost that simultaneously represents the caudal spot are par-

Female *Archocentrus septemfasciatus*

ticularly conspicuous. Even when all other bands are suppressed, they remain visible in the form of deep black markings. The ventral fins are largely black while all unpairy fins are blueish. The anterior section of the dorsal fin has metallic reflections. The iris is deep dark blue.

Male specimens lack the orange yellow shades, have no copper-coloured scales, and no black pattern in the dorsal fin that is, like the anal, long and pointed.

Specific traits

are preferably made up by the colour-pattern. Particularly characteristic features are the third band on the flanks that is often reduced to a lateral spot, and the bright blue colouration of the lower head and chest. Among the

similar species

within the genus *Archocentrus*, especially *A. spilurus* (GÜNTHER, 1862) deserves mentioning. This species, however, never shows a lateral spot. Moreover, its males have a loam-yellow coloured gular and chest region.

Natural habitats

of *Archocentrus septemfasciatus* are limited to the Atlantic slope of Central America. Its distribution range is relatively small and extends from southeastern Nicaragua to eastern Costa Rica. Localities are known from the drainage of the Rio San Juan and from the vicinity of Limon in Costa Rica. The home waters of this Cichlid usually have soft to moderately hard water with a pH ranging at slightly alkaline levels. As the

care

of this fish is very easy due to its adaptability and hardiness, it can also be recommended to aquarists with little experience in the keeping of Cichlids. An aquarium of

Catching fish in water bodies like this one is extremely difficult.

about a metre in length provides adequate space. It should by all means have some cave-like shelters among larger rocks and pieces of bog-oak. The bottom of the tank should be covered with sand or very fine gravel. More robust plants are ignored by the fish and may therefore serve as additional decoration items. As far as the chemical properties of the water and food are concerned, this Cichlid is absolutely undemanding.

Breeding

is therefore relatively easy. The fish preferably spawn in well covered places and make use of caves if such are available. The various tasks of parental care are subject to a strict task-allocation between the sexes. While the male is primarily responsible for guarding the breeding territory, the female dedicates herself to the care for the eggs and larvae. Since their parental care is very intense, breeding attempts are usually also successful in a community tank decorated according to nature.

57

▶ *Archocentrus spilurus*
(GÜNTHER, 1862)

Localities of *Archocentrus spilurus*

In its original description this Central American Cichlid of the tribe Cichlasomini was allocated to the genus *Heros*. The aquarium hobby knows it since the early seventies. Male specimens may reach a length of about 12 centimetres while females are considerably smaller.

On a pale grey to greyish blue background, the fish display seven dark bands between the edge of the gill-cover and the base of the caudal fin. The large caudal blotch covers almost the entire height of the caudal peduncle. Being not only dimorphous but also dichromatic, the sexes are easily distinguished.

With increasing age, the males lose the banded pattern until it has eventually vanished. Simultaneously the lower portion of the head, the throat, and the chest region assume a loam-yellow colouration. More-

Archocentrus spilurus; female caring for the brood

over, the forehead becomes increasingly bulged until the upper profile outline of the head eventually rises almost vertical from the mouth. Old specimens have enlarged and pointed dorsal and anal fins.

Female fish keep their banded pattern and it becomes even more pronounced during periods of parental care. They are furthermore characterized by a black spot in the dorsal fin that lies above the third lateral band between the ninth and eleventh spine. This marking is absent in males. During breeding, the lower head, the chest, and the belly are deep black.

Specific traits

are the pattern of seven equally shaped, dark lateral bands, the dark spot on the dorsal fin of the females, and the yellowish coloured head and chest of the males. This is complemented by 8 to 10 spines in the anal fin. A number of

similar species

are found in the genus *Archocentrus*. For example, *Archocentrus spinosissimus* (VAILLANT & PELLEGRIN, 1902) has a similar colour-pattern, but lacks the yellow colours in the males. This species also differs in having 11 to 12 spines in the anal fin. *Archocentrus septemfasciatus* (REGAN, 1908) also lacks yellow colours in the males. *Archocentrus sajica* (BUSSING, 1974) has a lateral stripe in addition to the lateral bands. The

natural habitats

of *Archocentrus spilurus* lie in Guatemala, Belize, and on the eastern border of the Mexican state of Campeche. Localities have especially been recorded from the Belize River, Rio Motagua, and the vicinity of the Laguna de Izabal. Further surveys are necessary to establish the exact borders of its range. Literature records from farther east, e.g. Nicaragua, Costa Rica, and

Archocentrus spilurus, male

Panama, refer to the species *Archocentrus septemfasciatus* (REGAN, 1908) and *Archocentrus cutteri* (FOWLER, 1932) which some authors considered synonyms of *Archocentrus spilurus*. The

care

of this Cichlid can also be recommended to the beginner as the fish is undemanding and very adaptable. This adds to the fact that it can also be successfully kept in tanks smaller than one metre in length. The chemical attributes of the water are of minor importance. Cave-like shelters contribute positively to the well-being of the fish. As they excavate shallow pits for their offspring, the bottom should be covered with sand or very fine gravel. There are no problems in planting the aquarium. In a tank set up as described before,

breeding

is relatively easy. *Archocentrus spilurus* is an openbrooder with a strong tendency towards cavebrooding. This means the fish willingly spawn on flat stones or similar substrates, but prefer to do this at well-hidden sites and readily choose a cave if such is available. The brood is cared for in a father-mother-family. Both parental specimens care for the offspring, but have different tasks. The male attends less to the eggs or the larvae, but rather protects the borders of the breeding territory. As the parental care is very intense, a few young fish also survive in a community tank.

The Genus Astronotus
SWAINSON, 1839

Together with the genera *Chaetobranchus* and *Chaetobranchopsis* this genus forms the tribe of the Chaetobranchines which differ from other genera by the possession of particularly large, strongly spinose micro-gill-rake. The extensively scale-sheathed bases of the dorsal and anal fin are particularly typical of the *Astronotus* species. Moreover, they are unusual with regard to the arrange-

ment of the lips with the posterior portion of the lower lip not covering the upper counterpart. Although a number of species were described in the past, the genus was temporarily considered monotypical. Today it is widely accepted that the genus incorporates several species of which, however, only *Astronotus ocellatus* from the upper Amazon River, and *A. crassipinnis* from the Madre de Dios, Mamoré, Guaporé, and Rio Paraguay have been defined (KULLANDER 1986).

The harbour of Manaus (Brazil) is the destination of many collectors of ornamental fish.

▶ *Astronotus ocellatus*
(AGASSIZ, 1831)

Localities of *Astronotus ocellatus*

originally had *Lobotes* for its generic name. Junior synonyms include, among others, *Acara compressus* COPE, 1872, and *A. hyposticta* COPE, 1878. It is also likely that *Cychla rubroocellata* JARDINE, 1843, is synonymous.

These fish may reach a length of 30 cm and are therefore an important source of food in their home countries. Their colour-pattern is dominated by diffuse brown to black bands that are in contrast with the grey or yellowish background colour. The scientific name refers to a round black spot on the upper half of the caudal peduncle that is surrounded by an orange red aura and a number of ocelli along the base of the dorsal fin. Individual orange red scales or colour-zones may be present on the flanks.

The species inhabits a vast area and it is therefore no surprise that various colour morphs are distinguished which mainly differ with regard to the extent of their orange red colour-zones. Selective breeding has produced a uniformly copper coloured strain that is referred to as "Red Oscar" by aquarists. This species is monomorphous, i.e. there are no external sexual differences.

Astronotus ocellatus has conspicuous spots on the dorsal fin.

Astronotus sp.

Specific traits

of *Astronotus ocellatus* are the conspicuous ocelli with their orange red surroundings as well as 12-14 spines and 19-21 rays in the dorsal fin. *A. crassipinnis* in contrast has no such ocelli and 11-13 spines and 19-22 rays in the dorsal.

Natural habitats

of *Astronotus ocellatus* are, according to our observations, primarily characterized by a slow flow, such as backwaters and bays. As we established in Peru, the fish prefer to stay in the cover of submerged branches and logs.

Following KULLANDER (1986) the distribution of *Astronotus ocellatus* extends over the river systems of the Rio Ucayali and the drainage of the upper Amazon River in Peru and Brazil. Here, localities are known from the vicinity of the towns of Iquitos, Pebas, and Codajás, and also from Rio Tefé and Rio Icá. It is likely that the lower Amazon is inhabited by another species. From this region, *Astronotus zebra* PELLEGRIN, 1904 and *A. orbicularis* HASEMAN, 1991, were described. Due to the size of this Cichlid, its long-term

care

requires very large aquaria that should at least measure one and a half metres in length. Attention must be given to the fact that specimens other than a bonded pair can be extremely aggressive towards each other if there is not sufficient space to avoid clashes. The decoration of the tank calls for gravel, larger stone-flakes or pieces of bog-oak, and some hardy plants to create hiding-places. *A. ocellatus* is quite hardy and undemanding so that the chemical properties of the water need no particular attention. Oscars are unaggressive towards other Cichlids of their own size and older specimens are very placid aquarium fish. They feed on large prey items and even semiadult specimens already need large amounts of food. They should therefore be regularly fed with pieces of lean fresh and sea-water fish.

A compatible pair in an aquarium of at least 120 centimetres in length can be conditioned for

breeding.

This Cichlid is an openbrooder which lies its eggs on a flat rock and cares for its brood in a parental family. In the case of larger specimens a clutch may contain a couple of hundreds of eggs. Provided with the usual diet, the juveniles grow rapidly. They have a pretty pattern of black and silver and look completely different from their parents.

Table 11:

Site:	Small white water stream in the upper Ucayali drainage near Iparia (Peru)
Clarity:	very turbid
Colour:	loam-yellow
pH:	7.3
Total hardness:	10 °dH
Carbonate hardness:	11 °dH
Conductivity:	455 µS at 28 °C
Depth:	3 m
Current:	slight
Water temperature:	28 °C
Air temperature:	30 °C
Date:	10.7.1981
Time:	12.30 h

The Genus Biotodoma

was described by Eigenmann & Kennedy in 1903. Its members are in possession of an epibranchial lobe, i. e. a flat expansion of the upper portion of the first gill-arch and the genus thus belongs to the Geophagines. It is distinguished from closely allied genera by the structure of the skeleton of the dorsal fin that contains two supraneurals (bones in the anterior dorsal fin skeleton). A trait usable in live fish is the position of the lateral spot that is never situated in the centre of the flank, but always in the posterior half. Varying with the species, it may either lie immediately below the dorsal fin on the upper lateral line, or between the lateral lines. At present, this small unit only consists of two species that will probably be complemented by another one in due course.

The natural habitats of the species of *Biotodoma* include black water biotopes like this water body in the Rio Negro.

◗ *Biotodoma cupido*
(HECKEL, 1840)

was assigned to the genus *Geophagus* when it was originally described. GÜNTHER (1862) then included it in the genus *Mesops*. Due to the rarity of imports, this otherwise highly recommendable species is seldomly kept in an aquarium. On reaching 15 cm it is fully grown.

The most significant characteristics in its colour-pattern are a well-developed black mask, a roundish black lateral spot bordered goldenly which lies between the base of the dorsal fin and the upper lateral line, and an orange to reddish brown colour-zone that may extend from the edge of the gill-cover up to the centre of the body in adult specimens.

A sexual dimorphism is weakly developed. Old males usually have larger fins and the outer rays of the caudal fin are enlarged by filaments.

Localities of *Biotodoma cupido*

Specific traits

are such as the lateral spot that lies between the base of the dorsal and the upper lateral line in the posterior portion of the body, and the orange brown colour-zone in the anterior part of the body.

Biotodoma cupido

Similar species

are *Biotodoma wavrini* (GOSSE, 1963) which is more slender in shape and whose lateral spot lies below the upper lateral line, and a form yet to be described from the Guyanas. Its lateral spot is situated on the upper lateral line. To date, it is included in the species *Biotodoma cupido*.

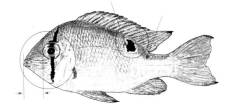

Biotodoma cupido

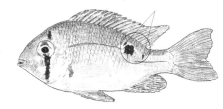

Biotodoma sp.

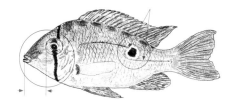

Biotodoma wavrini

The

natural habitat

of this Cichlid stretches from Brazil (Rio Solimoes, Rio Negro, Madeira, Araguaia, Tapajos), and northeastern Peru (Ucayali), to the border region between Bolivia and Brazil (Guaporé). Records from Guyana (Essequibo, Demerara) probably refer to another species of *Biotodoma*. This distributional pattern suggests that the fish generally occurs in soft water with a pH at acidic levels. Being a relatively small and very timid fish, its

care

can also be successful in comparatively small tanks of about one metre in length. The aquarium must be decorated with pieces of bog-oak and some thickets of plants in such a manner that an adequate number of covered zones and hiding-places are created. The bottom should preferably be covered with sand. This Cichlid is not really voracious, but nevertheless requires a nutritious and varying diet. It should include small crustaceans, White Mosquito larvae, and shrimp-meat. It is a peaceful, sometimes even shy species which can only be kept together with Cichlids of similar disposition, e.g. with Dwarf-cichlids.

Breeding

has only been successful in a very few cases. The first detailed report on the reproduction of a species of *Biotodoma* was published by KUHLMANN (1984). His specimens spawned in fairly hard water of 18° total and 9°dH carbonate hardness. It would certainly be of advantage to keep them in water poor in minerals with a pH at acidic levels. The fish spawned in a small pit that had been dug by the female. Both parents participated in an intense caring for the brood.

Table 12:

Site:	Quebrada de Bagasan in the lower Ucayali drainage (Peru)
Clarity:	turbid
Colour:	loam-yellow
pH:	6.5
Total hardness:	< 1°dH
Carbonate hardness:	3°dH
Conductivity:	85 µS/cm
Depth:	> 2 m
Current:	minor
Water temperature:	27°C
Air temperature:	25°C
Date:	1.8.1990

The Genus Bujurquina
KULLANDER, 1986

Incorporating some 30 species — only about 20 of which have so far been dealt with taxonomically (comp. KULLANDER 1994) — *Bujurquina* represents one of the largest clades of Cichlids in South America. The distribution of this genus ranges from northern South America (Colombia, Ecuador, Venezuela) to the river systems of the Rio Paraná in Argentina in the south. A particularly diverse variety of forms are found in Peru where altogether thirteen species have been recorded, eight of which occurring in the drainage of the Rio Ucayali.

The species of *Bujurquina* belong to the tribe of the Cichlasomines. Until 1986 they were considered representatives of the genus *Aequidens*. A closer relationship exists with the species of the *"Aequidens" pulcher* species-group and, on the other hand, with the genus *Tahuantinsuyoa* KULLANDER, 1986. Most species of *Bujurquina* resemble each other so closely that the identification of live fish is often extremely difficult, even when exact locality information exists.

A colour-feature found in all *Bujurquina* species is a black lateral stripe that obliquely runs from the edge of the gill-cover upwards to the posterior base of the dorsal fin, and, on the other hand, curves forward onto the nape passing along the posterior edge of the eye. Another common trait is the peculiar reproduction biology. All species of *Bujurquina* are larvophile mouthbrooders, which means they deposit their sexual products on a solid surface in the fashion of openbrooders. There, the clutch is cared for by both male and female until the larvae hatch. These are then taken into the mouth where they are kept until the larval development is completed.

This site on the Rio Cuieiras (Rio Negro drainage) is the locality for Cichlids of the genera *Apistogramma, Acarichthys, Aequidens, Cichla, Heros, Mesonauta,* and *Satanoperca.*

▶ *Bujurquina robusta*
KULLANDER, 1986

Localities of *Bujurquina robusta*

The specimen figured here was caught by ourselves during an ichthyological collecting trip to the uppe Ucayali drainage in 1983. Collaborators of ours brought several specimens back to Germany alive. KULLANDER examined the conserved material we had brought back and concluded that this was an undescribed species. The largest specimens we observed in the wild, measured about 15 centimetres. Females remain smaller.

The most notable black components of this species' pattern include a caudal spot on the upper half of the caudal peduncle and a lateral spot below the lateral line in the centre of the flanks. A black stripe begins on the nape, runs past the hind edge of the eye, and curves up to the upper anterior margin of the caudal peduncle. Depending on the mood, the side of the body may show six dark bands between the caudal spot and the gill-cover. The dorsal fin has a narrow orange yellow margin. The flanks have pretty metallic reflections that entail five blueish green lateral stripes. Similar glossy lines are visible below the eye in the area of the anterior head. Male and female fish are very difficult to classify on the basis of external characters. Older males eventually grow larger fins and their dorsal and anal fins are more produced.

Bujurquina robusta

Specific traits

Except the conspicuous metallic green sheen that is, in comparison with other Cichlids of this cladistic assembly, very intense in colour, *Bujurquina robusta* does not display specific features that would allow a separation of live fish from closely allied species. As the distribution ranges of *Bujurquina robusta* and *Bujurquina labiosa* overlap, it is of particular interest of how to tell these two species apart. *Bujurquina labiosa* has a much more pointed head and considerably thicker lips.

The

natural habitat

of this *Bujurquina* species lies on the upper Ucayali in Peru. The only locality known to date is the Rio Chinipo that flows into the Ucayali some 10 km south of the village of Chicosa. It is a clear water creek with a considerable current. The Cichlids, however, did not reside in the main channel, but in quiet backwaters and isolated pools where the bottom consisted of sand instead of rocks. Here, submerged logs and branches provided the necessary cover. The water was relatively soft and of distinct alkaline quality. Further Cichlids in these water-bodies were *Bujurquina labiosa*, *Cichlasoma amazonarum*, and *Crenicichla sedentaria*, which were accompanied by Sucker Catfishes of the genera *Chaetostoma* and *Hypostomus*, Characins of the genera *Phenacogaster*, *Bryconamericus*, and also *Hemigrammus agulha*, *Characidium etheostoma*, *Astyanax maximus*, and *A. bimaculatus*. Moreover, we found a Freshwater ray.

Tanks with a length of about one metre or more are appropriate for the

care

of this Cichlid. They should be decorated with sand or fine gravel, a few large rocks,

and pieces of bog-oak. An availability of hiding-places obviously contributes to the well-being of the fish. Although the natural biotopes did not contain higher plants, a sparse vegetation in the aquarium is recommendable, especially as this species tolerates plants. The tank can also be co-inhabited by smaller South American Cichlids, as this fish is relatively peaceful. It feeds on all the usual types of food.

Breeding

often needs a few attempts. It appears that partners cannot be united randomly. This Cichlid belongs to the larvophile mouth-brooders, which means the pair spawns on a stone or a root in the fashion of open-brooders. When the larvae hatch, they are taken into the mouth. Underwater observations revealed that this species forms a parental family-structure where both male and female take the fry back into the mouth when threatened. This observation was subsequently confirmed by POST who managed first to breed this species in captivity in 1984. At a water-temperature of about 27 °C, mouthbrooding begins after some 40 hours. The juveniles leave the mouths of the parents after round about one week.

Table 13:

Site:	Isolated rest-water pool on the Rio Chinipo (clear water river), about 10 km south of Chicosa (Depto. Junin, Peru)
Clarity:	3 m
Colour:	greenish
pH:	7.9
Total hardness:	4 °dH
Carbonate hardness:	4 °dH
Conductivity:	116 µS at 26 °C
Depth:	< 1 m
Current:	none
Water temperature:	26 °C
Air temperature:	30 °C
Date:	23. 6. 1983
Time:	11.00 h

◗ *Bujurquina tambopatae*
KULLANDER, 1986

This Cichlid unquestionably ranks among the most splendid representatives of this genus. It has a greyish to loam-yellow background colouration overlain on the flanks with a marvellous metallic light green sheen in adult specimens. The lower head region is also marked with shiny green dots and streaks. The anterior portion of the dorsal fin and the upper half of the caudal fin have a narrow marginal stripe of either orange red or white colour. This variation is even present in specimens from one and the same locality. All fins have a pale yellowish hue with the unpairy ones showing a pattern of small pale blue dots in certain light situations.

The intensity of black markings depends on the prevailing mood, but they generally consist of those typical of the genus. The lateral stripe is usually broken into a series of blotches. It runs from the upper edge of the gill-cover up to the posterior end of the dorsal fin and, on the other hand, curves over the nape region. The space between the

Localities of *Bujurquina tambopatae*

posterior margin of the gill-cover and the base of the tail is marked with seven bands, added by a small elongate caudal spot.

The maximum total length of this species is about twelve centimetres. As there is no distinct sexual dimorphism, the sex of a specimen is difficult to determine on the basis of external characters. Old males, however, have longer ventral fins. Their dorsal and anal fins are more elongate and

Bujurquina tambopatae; specimen from the type locality

filamentous. Moreover, their dorsal fins have no smooth but fray margins, a trait that is characteristic of many species of *Bujurquina*.

Specific traits

Due to the fact that the distribution areas of *Bujurquina tambopatae* and *Bujurquina cordemadi* overlap, it is of particular interest which specific features distinguish these two species. In the case of *B. cordemadi*, the fifth and sixth lateral band — counting from back to front — are fused dorsally while they are separated in *B. tambopatae*. *B. tambopatae* furthermore has thicker lips and a longer muzzle.

Natural habitats

This Cichlid is only known from its type locality. It is a small, in its upper section creek-like river named Quebrada San Roque or alternatively Quebrada Castañas. It crosses the road leading from the town of Puerto Maldonado to Cuzco approximately ten kilometres south of the former. This stream flows into the Rio Tambopata that itself is a tributary to the Madre de Dios.

In August 1992, we could find several adult specimens of *Bujurquina tambopatae* in a pond-like backwash of this waterbody. An analysis of the fairly turbid water revealed that it was very soft and slightly acidic. This locality was surprisingly rich in species of fish including the Killifish *Rivulus christinae* and the Cichlids *Mesonauta festivus*, *Cichlasoma boliviense*, *Aequidens tetramerus*, *Crenicichla semicincta*, and *Apistogramma urteagai*.

Care

The care of this Cichlid requires an aquarium of at least one metre in length. It may be planted with *Echinodorus* and larger species of *Anubias* as these will not be damaged by the fish. These plants must, however, not unduly limit the available swimming space. The decoration of the tank according to nature furthermore entails a few larger rocks and roots among which there are hiding-places. As *Bujurquina tambopatae* is not a bold species, it should not be kept in the company of larger or aggressive species.

Breeding

Bujurquina tambopatae belongs to the larvophile mouthbrooders. The eggs are attached to a solid surface much like in the fashion of openbrooders where they remain until the larvae hatch. Mouthbrooding begins only then, and both male and female participate in the form of a parental family.

For breeding attempts it is advisable to emulate the water-values of the natural habitat as closely as possible. In hard water with a distinctly alkaline reaction there is a good chance of the larvae not developing normally. When the young fish leave the mouths of their parents, they are already large enough to feed on freshly hatched larvae of the Brine shrimp.

Table 14:

Site:	Quebrada San Roque/Castañas in the Tambopata drainage, on the road to Cuzco some 10 km from the town of Puerto Maldonado
Clarity:	poor
Colour:	loam-yellow
pH:	6.7
Total hardness:	< 1 °dH
Carbonate hardness:	< 1 °dH
Conductivity:	10 μS/cm
Depth:	> 1 m
Current:	slow
Water temperature:	21.5 °C
Air temperature:	26.0 °C
Date:	3. 8. 1992
Time:	15.30 h

The Genus Caquetaia
FOWLER, 1945

This genus was suggested by FOWLER, on occasion of the original description of *Caquetaia amploris*, a taxon that meanwhile has turned out to be a synonym of *Caquetaia myersi*. This genus then was temporarily dumped into the synonymy of *Cichlasoma*, but was revalidated by the Swedish ichthyologist KULLANDER in 1986 when he undertook to redefine and split the former composite unit *Cichlasoma*. That the name *Caquetaia* was available for a clade of South American Cichlids was pointed out by KULLANDER as early as in 1983.

The genus *Caquetaia* belongs to the Cichlasomines and presently accommodates only three large South American Cichlid fishes. All these are highly specialized predatory piscivorous fishes. They share the trait of a large, very stretchable mouth with enlarged teeth in the front portion. Their fins are scaly, the lateral scales are small, and they have five to seven spines in their anal fins. Close relationships exist with the genus *Petenia* which is endemic to Central America. The species of *Caquetaia* are native to Colombia and the Amazon basin.

▶ Caquetaia myersi
(SCHULTZ, 1944)

In its original description, this Cichlid was assigned to the genus *Petenia*. A junior synonym is *Caquetaia amploris* FOWLER, 1945.

Caquetaia myersi has a very deep, laterally strongly compressed body and a large, stretchable mouth. Its basic colour is a dirty greyish yellow that may change into a bright loam-yellow if the fish becomes excited. Individual scales on the flanks have iridescent reflections. The iris is coloured deep red. All unpairy fins usually have a pale blue hue.

Localities of *Caquetaia myersi*

The intensity of black markings largely depends on a variety of stages of excitement. Besides the super- and suborbital stripe, there are normally just two more black bands clearly visible. The anterior one lies in the centre of the body while the posterior one links the posteriormost spines of the dorsal and anal fins. These may be complemented by another band between these two and even another three on the hind portion of the body, with the posteriormost band then covering the base of the caudal fin.

Identifying the sexes of semi-adult specimens is difficult as there is no pronounced sexual dimorphism. In old specimens, the fins provide usable hints as they become more expansive in the males. These also grow larger than the females. *Caquetaia myersi* may reach a length over 25 centimetres.

Natural habitats

The type locality of *Caquetaia myersi* lies on the Rio Dedo in Colombia, a tributary to the Rio Orteguaza. More sites are known from the drainage of the Rio Caqueta. Furthermore the species also occurs in the

Pair of *Cacquetaia myersi* caring for their clutch

upper drainage of the Amazon River in Ecuador, including the river systems of the Rio Aguarico, Rio Napo, and Rio Coca. Ingemann HANSEN (in litt.) made available to us the results of measurements he took in the wider vicinity of the town of Florencia (Caqueta Province) on localities where *Caquetaia myersi* is found in the drainage of the Rio Orteguaza. These include streams like the Quebrada Mochilero, Quebrada Aguas Calientes, and Quebrada Montanita. According to this information the Cichlid inhabits very soft water that often has an alkaline reaction. Carbonate hardness was always below the detectability- level of 1 °dH, while the total hardness did never exceed 2 °dH. The pH-value varied between 6.0 and 8.5. At air temperatures between 29 and 33 °C, those of the water measured 24 to 30.5 °C.

Care

Caquetaia myersi is a large Cichlid which grows fairly rapidly if given ample of food.

Its long-term care can therefore be only successful in very spacious, custom-made aquaria of at least one and a half metres in length. On the other hand, this fish is calm and fairly inactive. The tank should be decorated with rocks and roots so that a number of caves and shelters are created and borders of territories are defined. The use of robust plants is optional.

Being a predatory piscivorous fish, this Cichlid requires rich food such as meat of fish and (deep-sea) shrimp.

Breeding

For breeding attempts it is recommendable to keep a compatible pair on its own in a separate tank that is decorated according to nature. *Caquetaia myersi* is a typical open-brooder which deposits its eggs on a solid surface. Both male and female care for the eggs, larvae, and young fish in a parental family. Newly hatched nauplii of the Brine shrimp have proven adequate for initial feedings.

Chaetobranchopsis australis lives in lagoons and small streams that often carry white water.

The Genus Chaetobranchopsis
STEINDACHNER, 1875

belongs to the tribe Chaetobranchini. It is a small genus containing only two species of minor importance for the aquarium hobby. A close relationship exists with the genus

Chaetobranchus HECKEL, 1840. Both differ from other American Cichlids, by number and the length of the gill-rakers on the lower section of the first gill-arch. While *Chaetobranchus* has only three spines in the anal fin, the genus *Chaetobranchopsis* possesses four to six.

73

▶ *Chaetobranchopsis australis*
EIGENMANN & WARD, 1907

Localities of *Chaetobranchopsis australis*

This Cichlid has only made a few appearances in the aquarium hobby as it occurs in regions where ornamental fishes are rarely exported from.

Chaetobranchopsis australis may reach a length of 20 cm. Its flanks have a metallic silvery to brass sheen. The iris is blood-red and the lower portion of the head is yellowish to bright yellow. Approximately in the centre of the body there is a larger black spot that touches the upper lateral line with its upper zenith. The upper angle of the gill-cover holds a reddish to brownish red spot. The central scales of the first six scale rows bordering the gill-cover may also be spotted with red.

As there is no distinct sexual dimorphism, distinguishing between the sexes is difficult.

The most significant

specific trait

that allows an identification of this Cichlid is the number of spines in the anal fin that normally count five.

Chaetobranchopsis australis

Among the

similar species

with more or less the same body shape the species of *Chaetobranchus* deserve mentioning. They, however, have only three anal spines.

Chaetobranchopsis orbicularis STEINDACHNER, 1875, usually has six spines in the anal fin and is easily recognized by the presence of two dark stripes on the flanks.

The

natural habitat

of this Cichlid appears to centre in the border region between southern Bolivia, northern Paraguay, and Brazil. Literature records published so far all lie in the drainage of the upper Rio Paraguay in the vicinity of the villages of Bahia Negro and Puerto Suarez and on the upper Jaura near Campos Alegre. The authors managed first to record this species from farther west in 1984, i.e. from central Bolivia. Our locality lies in the drainage of the Rio Marmoré in the vicinity of the town Trinidad. The fish occurred there in pond-like expansions of small streams and lagoons that are common on either side of the cross-country roads. These are typical white water biotopes with

Table 15:

Site:	Rest-water in the Rio Marmoré on the road from Trinidad to Puerto Amacen (Depto. Beni, Bolivia)
Clarity:	extremely turbid
Colour:	whitish
pH:	6.6
Total hardness:	1.5 °dH
Carbonate hardness:	1.0 °dH
Conductivity:	37 µS at 29 °C
Depth:	> 2 m
Current:	none
Water temperature:	29 °C
Air temperature:	31 °C
Date:	11. 7. 1983
Time:	15.00 h

extraordinary turbidity, but the water is very soft and acidic. Except one species of *Utricularia*, there were no other aquatic plants. The water surface though was densely covered with a carpet of floating plants, the majority of which being *Eichhornia crassipes*. Among other fishes recorded from these sites were the Cichlids *Cichlasoma boliviense*, *Crenicichla lepidota*, *Apistogramma staecki*, the Characins *Hoplias malabaricus*, *Pyrrhulina brevis*, *Charax gibbosus*, *Prinobrama filigera*, *Ctenobrycon spilurus hauxwellianus*, the Knife-fishes *Gymnotus carapo*, *Eigenmannia* sp. aff. *virescens*, *Hypopomus* sp., and the Catfishes *Parauchenipterus galeatus* and *Hoplosternum littorale*. It was noteworthy that the collecting sites were also home to huge numbers of small freshwater shrimp.

The long-term

care

of this Cichlid will only be possible in an aquarium of more than one metre in length. When decorating the tank attention should be given to the fact that the fish need to find hiding-places among sensibly arranged pieces of bog-oak and rock-flakes. Although the high number of long, slender gill-rakers that are typical of all *Chaetobranchus* suggests that these fishes feed at least temporarily on very small prey, the observations made in their natural biotopes indicate that freshwater shrimp play an important role in their diet. In the aquarium they should therefore be occasionally given cut pieces of shrimp meat.

No information is yet available on the

breeding

of this species under aquarium conditions. Acidic water poor in minerals appears to be an ideal choice for those attempts. It should furthermore be taken into consideration that the local water-temperature was 29 °C.

The Genus Chuco
FERNANDEZ-YEPEZ, 1969

The name *Chuco* was long considered a synonym of *Cichlasoma* or *Theraps* respectively. That it is available for a clade of Central American Cichlids was pointed out by KULLANDER in 1983 when he revised the genus *Cichlasoma*, though he failed to formally implement the name *Chuco* in its new function. This step was eventually taken by ALLGAYER in 1989 who explicitly referred to the generic criteria as lined out by FERNANDEZ-YEPEZ. At present, this genus accommodates three species.

▶ *Chuco intermedius*
(GÜNTHER, 1862)

was assigned to the genus *Heros* when originally described. Synonyms of this species include *Heros angulifer* GÜNTHER, 1862, and *Acara rectangularis* STEINDACHNER, 1864. The fish may grow up to a length close to 20 cm. Larger numbers were first imported into Europe in 1983, amongst others by ourselves. Breeding was first successful in 1984.

This species is characterized by a greyish brown background colouration that assumes a rather yellowish brown tint on the

Localities of *Chuco intermedius*

upper half of the body. A black angulate pattern consisting of a broad lateral stripe that begins on the rear edge of the eye and ends at mid-body at approximately the level of the thirteenth dorsal spine is a very typical feature. There it connects with the fifth lateral band that comes down from above, but also ends in the centre of the flank. The remaining eight lateral bands are reduced to indications or are absent altogether. The caudal blotch is distinct though. The green iris is conspicuous.

During periods of parental care, the chest and belly areas assume a black colouration, and the angulate pattern and parts of the bands stand out more contrasting. As the species is isomorphic, males and females cannot be distinguished with certainty on the basis of external characters alone.

Specific traits

that may help to identify this species are the angulate black pattern, the very short ventral fins, and the low number of just 6 to 8 gill-rakes on the lower branch of the first gill-arch. Since the other two species of *Chuco* have completely different colourpatterns they can hardly be mistaken for *C. intermedius*.

Chuco intermedius in its natural biotope

Chuco intermedius

Natural Habitat

The distribution of this species spans over the northwestern parts of Guatemala and the Mexican states of Tabasco and Chiapas. Localities were preferably recorded from the drainage of the Rio Usumacinta and Rio Grijalva. We repeatedly found the fish in the vicinity of Palenque where they inhabited typical clear water streams.

Care

Problems in the captive care of this Cichlid are sometimes encountered in form of single specimens becoming fairly aggressive towards each other. It is therefore recommendable to keep them in spacious tanks of about one and a half metres. The decoration of the aquarium first and foremost requires larger rocks that are arranged to form hiding-places. These may be complemented by pieces of bog-oak and even plants as this Cichlid is plant-tolerant. Suitable co-inhabitants of the tank are preferably found among other Mexican Cichlids. This fish readily feeds on flake-food and other common types of food. Larger specimens obviously require a nutritious diet. Provided with enough space

breeding

can also be successful in a community tank. The fish spawn in the fashion of open-brooders and favour a stone at a covered site. Both male and female participate in caring for and protecting their brood. Rearing the juveniles on the larvae of the Brine shrimp is not difficult.

Table 16:

Site:	Rio Nututun near Palenque (Chiapas, Mexiko)
Clarity:	ca. 3 m
Colour:	greenish
pH:	8.85
Total hardness:	12.5 °dH
Carbonate hardness:	12.0 °dH
Conductivity:	1200 µS at 27 °C
Depth:	150 – 350 cm
Current:	strong
Water temperature:	27 °C
Air temperature:	30 °C
Date:	2. 4. 1983, 14.00 h

The Genus Cichla
SCHNEIDER, 1801

Although not less than fifteen species of *Cichla* were described over the years, the genus had been considered to be a very small clade consisting of just two or three species with the other taxa being synonyms. Nowadays (KULLANDER & NIJSSEN 1989) it has been widely accepted that there are at least eleven valid species of *Cichla*. Of these, only five have been dealt with in recent years by MACHADO-ALLISON (1971, 1973), KULLANDER (1986), and KULLANDER & NIJSSEN (1989). These are *Cichla intermedia* MACHADO-ALLISON, *C. monoculus* SPIX, *C. ocellaris* SCHNEIDER, *C. orinocensis* HUMBOLDT, and *C. temensis* HUMBOLDT.

All species of *Cichla* proceed through various age-dependent stages of colour-pattern which makes it extraordinarily difficult to identify live juvenile or semi-adult specimens. Fully grown specimens of the five species mentioned above are easily recognized on the basis of specific colour traits.

Phylogenetically, *Cichla* is considered to represent a very primitive clade. Certain features of its members correspond much better with those found in African than in any other American Cichlids. These particulars include the morphology of the lips that are in so far noteworthy as the posterior portion of the lower lip does not cover its counterpart of the upper lip like in most other South American Cichlids.

The representatives of the genus *Cichla* rank among the largest Cichlids. According to reports by anglers, they may exceed 60 centimetres in length and weigh three to four kilograms. It is therefore of little surprise that they belong to the favourite fish on Amazonian fish-markets.

They are readily distinguished from all other American Cichlids by the typical shape of the dorsal fin, which shows a deep cleft in front of its rayed portion. More

Localities of *Cichla monoculus*

characteristic traits include the deeply split mouth with the protruding lower jaw and the scaly fins.

▶ *Cichla monoculus*
(SPIX, 1831)

Adult specimens of *Cichla monoculus* exhibit a yellowish green to greenish yellow basic colouration on the flanks. The lower parts of the body are white. The transitory area between these two colour-zones is often marked with a narrow bright orange zone that extends from the angle of the mouth to the lower base of the caudal fin. The iris is orange red and the lower portion of the caudal and the anal fins are reddish brown. The flanks are marked with three broad deep black bands that often end at mid-body. The lower half of the flanks may be patterned with more black markings in the form of spots. The upper base of the caudal fin holds a conspicuous ocellate blotch with a golden aura.

This Cichlid lacks clear external sexual traits. Sexually mature males may grow a fat bulge on the nape though.

Cichla monoculus shortly after being caught in the Rio Ucayali

Specific traits

Characteristic colour features that allow an identification of adult specimens of *Cichla monoculus* include the dark pattern on the flanks. These consist of three black bands that, in contrast to other species of *Cichla*, begin immediately below the dorsal fin and lack shiny golden or silvery margins.

Natural habitats

Localities of *Cichla monoculus* are mainly known from the upper drainage of the Amazon River in Ecuador, Peru, and Brazil. This fish has been recorded from the central and lower Rio Ucayali as well as from the lower Rio Napo and the Rio Tefé. It is said it would occur in the Rio Solimoes up to Codajás or even as far down as to Manaus. KULLANDER (1986) also indicated the river system of the Madre de Dios in Bolivia (Rio Guaporé, Rio Beni) and the Oyapock in Guyana as possible localities.

Care

Due to its considerable adult size this Cichlid is certainly no fish for an ordinary aqua-rium in the living room. It is occasionally offered as a juvenile, but its care must be restricted to those aquarists who possess really large tanks that guarantee an adequate long-term housing.

Cichla monoculus is a predatory pisci-vorous fish which often will initially feed on live fish only and refuse all types of sub-stitute food. Attempts to get the fish to also accept dead food should start off with dead Smelts that are loosely attached to a thread and moved around.

Table 17:

Site:	Paca Cocha (Rio Ucayali drainage) in the vicinity of the town of Pucallpa in Peru
Clarity:	very turbid
Colour:	yellow
pH:	7.7
Total hardness:	14 °dH
Carbonate hardness:	16 °dH
Conductivity:	474 µS/cm
Depth:	> 2 m
Current:	hardly noticeable
Water temperature:	28.6 °C
Air temperature:	30.0 °C
Date:	31.7.1985
Time:	16.00 h

◆ *Cichla orinocensis*
HUMBOLDT, 1833

Localities of *Cichla orinocensis*

For decades, *Cichla orinocensis* was erroneously regarded a synonym of *Cichla ocellaris* until KULLANDER (1986) provided evidence that these two were separate species. The body of *Cichla orinocensis* is basically greenish yellow with a white ventral side. In the case of this impressive large Cichlid, the three dark bands that are present in the juveniles of all species of *Cichla* are reduced to three large, very irregular lateral blotches with golden surroundings in older specimens. Between these markings there may be more, but considerably smaller black spots. Another ocellate spot with a golden aura is found on the upper half of the base of the caudal fin. Its ventral and anal fins, and also the lower half of the caudal fin are orange red to reddish brown. Adult specimens have a conspicuous blood-red iris.

Cichla orinocensis may reach a total length in excess of 60 centimetres which is why this fish is sought after for food purposes. A sexual dimorphism appears to be absent in this Cichlid.

Cichla orinocensis

80

This species of *Cichla* from Manaus shows an unusual red colouration

Specific traits

As far as the colour-pattern is concerned, specific characteristics of adult specimens of *Cichla orinocensis* are such as the three large, black, irregularly shaped lateral blotches with their golden surroundings, the latter being differently coloured in related forms.

Natural habitats

Locality records of *Cichla orinocensis* primarily centre in Venezuela where this species occurs in the river system of the Orinoco as well as in the drainage of the upper Rio Negro. As to how far south this Cichlid occurs in the Orinoco, remains yet to be established.

Care

Due to its unusual adult size this Cichlid is no aquarium fish in the common sense though it makes an impressive display specimen in any public show tank. On the other hand, there are also private keepers who have aquaria of a size available that would allow a sensible keeping of this large Cichlid. *Cichla orinocensis* is a fairly calm fish which only shows its incredible agility when it comes to feeding with live fish. Being a piscivore, it is often difficult to adjust it to a substitute diet in the form of fish meat.

Localities of *Cichla temensis*

▶ *Cichla temensis*
HUMBOLDT, 1833

This Cichlid has a large, deeply split mouth and a body that is more slender and elongate than in the cases of some other species of *Cichla*. Its flanks have a grey to silvery white colouration with the flanks being marked with three narrow, deep black bands that extend from the base of the dorsal fin down to the lower sides. The first band lies shortly behind the pectoral fin, the second a small distance in front of the anus, and the third at the level of the first anal fin rays. Between the rear edge of the eye and the edge of the gill-cover there is a short horizontal postorbital stripe.

The upper half of the base of the caudal fin holds a conspicuous ocellate spot with a silvery to golden coloured aura. Occasionally this spot may appear as a double blotch that then extends onto the lower portion of the fin. The entire body is speckled with small spots and little dashes that glitter silvery or golden and that are arranged to form four to six irregular horizontal rows. The iris is golden to reddish, and the pectoral fins are pale orange. The ventral fins, the anal fin, and the lower half of the caudal have a

Cichla temensis

bright orange red shade. The upper half of the latter, and the dorsal fin, are patterned with small light speckles in several rows.

Specific traits

The colour features that allow a positive identification even in the case of a semi-adult *Cichla temensis* include the numerous small, shiny silvery to golden spots and dashes arranged in several irregular rows on the sides. Moreover, *Cichla temensis* has smaller scales than all other representatives of this genus resulting in the unusually high number of 108 to 127 scales in a horizontal row.

Natural habitats

Cichla temensis lives in Venezuela and in northern Brazil. Localities are known from the river systems of the Orinoco and Rio Negro. It is commonly encountered on the fish market of Manaus. We caught it repeatedly with an angle in the region of the

Anavilhanas Islands where the forested banks were flooded by highwater at that time.

Care

A sensible long-term keeping of these giant piscivorous predatory fish is only possible in special tanks.

Table 18:

Site:	Rio Cuieiras in the lower Rio Negro drainage (Brazil)
Clarity:	2 m
Colour:	dark brown
pH:	4.3
Total hardness:	< 1 °dH
Carbonate hardness:	< 1 °dH
Conductivity:	10 µS/cm
Depth:	> 2 m
Current:	not recognizable
Water temperature:	26 °C
Air temperature:	23 °C
Date:	25. 3. 1986
Time:	6.00 h

The Genus Cichlasoma
SWAINSON, 1839

Since its revision by REGAN in 1905, *Cichlasoma* represented a heterogeneous catch-all genus containing more than one hundred, mainly Central American species. It was then revised again by KULLANDER in 1983 and restricted to about a dozen species from South America with the previous Central American members being spread over a number of separate genera.

The genus is widely distributed in tropical and subtropical South America from the Orinoco drainage in the north up to the river system of the Paraná in the south. The representatives of this genus are moderately large Cichlids whose total length does not exceed 15 centimetres. Most of the species resemble each other so closely in shape and colour-pattern that the determination of live specimens is extremely difficult if there is no exact information on their individual origin. They all share the pattern features of a small cheek spot below the eye, a large blotch on the flanks, and a caudal spot on the upper half of the base of the caudal fin. This is complemented by a broad lateral stripe that runs from the eye to at least the lateral spot or extends farther to the caudal spot.

The clear water rivers of Mexico are the home of numerous Cichlids that were previously assigned to the genus *Cichlasoma*.

83

▶ Cichlasoma amazonarum
KULLANDER, 1983

Localities of *Cichlasoma amazonarum*

Until it was recognized as a separate species in 1983, this Cichlid was included in the species *Cichlasoma bimaculatum* (LINNAEUS, 1758). Specimens caught in the wild measure about 12 centimetres at maximum, but they can grow slightly larger in captivity.

This fish is not particularly colourful. The body is light greyish brown to brown with a pattern of blackish markings. This includes a caudal spot that is often enlarged to form a band, a large lateral spot that lies in the centre of the flank immediately below the upper lateral line, and a small cheek spot below the eye. There is also a black lateral stripe whose anterior portion between the eye and the lateral spot is usually much more pronounced than its posterior continuation. Eight lateral bands may be present as indications. The upper part of the iris is red. The sexes cannot be distinguished with certainty on the basis of external characters alone.

Specific traits

that make an identification of *Cichlasoma amazonarum* possible and differentiate it from similar species are such as a multitude of tiny scales on the base of the dorsal and anal fins and the absence of a clearly defined lateral stripe on the posterior portion of the body. It is furthermore important to note

Cichlasoma amazonarum

that this Cichlid usually has four spines in the anal fin although a slight variation may occasionally occur.

Similar species

that also have four anal spines include *Cichlasoma orientale* KULLANDER, 1983, which has a more distinct lateral stripe on the rear portion of the body, *C. sanctifranciscense* KULLANDER, 1983, with its unmarked caudal fin, and *C. orinocense* KULLANDER, 1983, as well as *C. taenia* (BENNET, 1831) which both have less extensively scaly bases of the dorsal and anal fins than *C. amazonarum*.

The natural habitats

of *C. amazonarum* are widely scattered over the Amazon region. Locality records exist for the entire Rio Ucayali in Peru, the Rio Solimoes, and also the Amazon River in Brazil. We found this Cichlid during several expeditions in the drainage of the central and upper Ucayali and on the Rio Tambo where it was the most common Cichlid on many collecting sites. Here it mainly inhabits white waters, but we also caught it in Rio Chinipo that is a typical clear water river. It appears to be characteristic of this species that it largely avoids flowing and deep water. Its preferred biotopes are shallow backwaters and isolated pools with ample branches and rotting leaves or that are densely vegetated. There are indications that it can also survive in relatively poorly oxygenated water. Analyses on a number of localities revealed a pH of 7.1 to 7.9. The total hardness varied between 4 and 13.5° and the carbonate hardness between 4 and 16.5 °dH. Water-temperatures were established to range from 26 to 29 °C. Despite *C. amazonarum* living in such a variety of waters, it is typically found in classical biotopes of *Apistogramma*-species and Killifishes. We often caught it alongside with

A. eunotus, A. cacatuoides, and Killifishes of the genera *Pterolebias* and *Rivulus*.

Care

As the observations made in the natural biotopes of this Cichlid already suggest, it is an adaptable and little sensitive aquarium fish that also the beginner can keep with success. A compatible pair will be happy in a tank of nearly a metre in length. Such a tank should be densely planted and decorated with larger roots and rocks as the fish are only comfortable if there is sufficient cover. They accept all commonly used types of food. Attention to the chemical properties of the water can largely be neglected as this is an adaptable species. Since it is not aggressive, it may be kept together with other fishes in a community tank. If this is not too crowded and leaves enough space,

breeding

will be successful. The fish is a typical open-brooder which spawns on solid surfaces. Both parents care for the brood, and the larvae are transferred to a well-covered ditch or pit. Young fish can initially be fed with pulverized flake-food and larvae of the Brine shrimp.

Table 19:

Site:	Rest-water on the Rio Chinipo (clear water river), about 10 km south of Chicosa (Depto. Junin, Peru)
Clarity:	3 m
Colour:	greenish
pH:	7.9
Total hardness:	4 °dH
Carbonate hardness:	4 °dH
Conductivity:	116 µS at 26 °C
Depth:	< 1 m
Current:	none
Water temperature:	26 °C
Air temperature:	30 °C
Date:	23. 6. 1983
Time:	11.00 h

The natural habitat of *Cichlasoma boliviense* north of the town of Santa Cruz (Bolivia). This Cichlid is highly adaptable and occurs naturally in very different environments, including very warm rest-waters.

◆ *Cichlasoma boliviense*
KULLANDER, 1983

Until this species was described, this fish was usually referred to as *Aequidens portalegrensis*. It may grow to a length of about 12 centimetres.

The most significant dark elements of the colour-pattern include a small spot on the upper margin of the base of the caudal fin or on the caudal peduncle, a lateral spot at approximately the level of the eighth to tenth dorsal spine immediately below the lateral line, and a cheek spot in the form of a curving streak on the hind or lower edge of the eye. Seven to eight lateral bands and one lateral stripe are reduced to indications. The posterior portion of the body often appears mingled with black as many scales carry a dark dot. The rear sections of the anal and dorsal fins and the entire caudal fin may

Localities of *Cichlasoma boliviense*

have pale yellow hues. The cheeks also shine yellowish. Being an isomorphic species, sexes are difficult to tell apart by external characters only.

Cichlasoma boliviense

A specific trait

of this fish is the fact that the bases of all unpairy fins are densely covered with tiny scales. Moreover, *C. boliviense* usually has only three spines in the anal fin. It is also important for the aquaristic praxis that this Cichlid shows a pattern of two or three narrow, vertical, black lines on the base of the caudal fin and that the posterior half of the body is marked with black dots.

Similar species

that also possess only three anal spines are *Cichlasoma araguaiense* KULLANDER, 1983 with its substantially enlarged cheek spot and an irregular spotted pattern on the gill-cover, *C. dimerus* (HECKEL, 1840) with its blueish green coloured head and flanks, and *C. portalegrensis* (HENSEL, 1870) which is characterized by only 11-15 teeth on the outer half side of the lower jaw. The

natural habitats

of *C. boliviense* lie in the Bolivian portion of the Amazon basin covering the Rio Madeira drainage. At certain localities this species is the most common Cichlid. In 1983, we repeatedly recorded this fish from the vicinity of Trinidad on the Rio Marmoré and near Santa Cruz on the Rio Japacani and Rio Grande. Analyses of the water at a number of collecting sites showed that *C. boliviense* naturally occurs in a wide variety of environmental conditions. The total hardness was recorded to range from 1 to 15 °dH, the carbonate hardness varied from 1 to 16.5 °dH, and the pH-value was established to lie between 5.7 and 7.8 with the water-temperatures ranging from 24 to 29.5 °C. Our observations appear to suggest that this Cichlid preferably lives in stagnant water-bodies that provide cover in the form of plant thickets. This condition was partly provided by submerged land vegetation, or by floating plants such as *Eichhornia*

crassipes and *E. azurea*, or by true aquatic plants *(Echinodorus)*. The fish was also present in very warm and oxygen-deficient, cutoff pools. It was found to occur syntopically with *Crenicichla lepidota, Bujurquina vittata, Laetacara dorsigera*, various species of *Apistogramma*, Catfishes such as *Rhineloricaria, Pimelodella*, and *Hoplosternum*, Knife-fishes of the genera *Eigenmannia* and *Hypopomus*, and Characins representing *Moenkhausia, Hemigrammus, Prionobrama, Cheirodon* and *Pyrrhulina*. As can be perceived from the observations made in the natural environment, *C. boliviense* is a robust and adaptable species so that its

care

does not create particular problems. The length of the aquarium should be about one metre. It should be decorated with sand and gravel, rocks, and roots. It is imperative to also include some thickets of aquatic plants that provide the same sort of cover present in the natural biotopes. Among each other and towards other Cichlids, this species is unaggressive. Their

breeding

can be successful under the most different of keeping conditions. They are typical openbrooders and both parents participate in caring for their offspring.

Table 20:

Site: Lagoons on the road from Portachuelo to Bella Vista (Dept. Santa Cruz, Bolivia)	
Clarity:	very turbid
Colour:	brownish
pH:	7.8
Total hardness:	6 °dH
Carbonate hardness:	6 °dH
Conductivity:	198 µS at 24 °C
Depth:	10 – 100 cm
Current:	hardly any
Water temperature:	24.0 °C
Air temperature:	30.5 °C
Date:	6. 7. 1983, 11.00 h

▶ Cichlasoma dimerus
(HECKEL, 1840)

This Cichlid was originally described as *Acara dimerus*. Temporarily, it was later erroneously referred to as *Aequidens portalegrensis* in the aquaristic as well as in the scientific literature. The traits that distinguish *C. dimerus* from allied representatives of this genus include blueish green reflections on the flanks of adult males, the heavily marked caudal fin, a very small cheek spot below the eye, and the low number of three spines in the anal fin.

This species has a fairly well developed sexual dimorphism with the males not only growing much larger fins with increasing age, but also having this splendid blueish green shine on the flanks. In contrast, the females are rather brownish.

Natural habitats

Locality records for *Cichlasoma dimerus* are known from the four countries Bolivia,

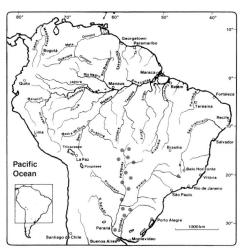

Localities of *Cichlasoma dimerus*

Brazil, Uruguay, and Argentina. KULLANDER (1983) indicated the distribution to cover the entire river system of the Rio Paraguay, the lower Alto Paraná, and the Paraná up to the vicinity of Buenos Aires. We recorded the fish from the Pantanal and also from the drainage of the Rio Cuiabá

Male of *Cichlasoma dimerus*

(Brazil), the vicinity of Puerto Suarez (Bolivia), and the system of the Paraná in Argentina. Accordingly, the fish does not only live in tropical, but also in subtropical conditions and therefore is surprisingly tolerant of low water-temperatures. A typical feature of their biotopes is a profuse vegetation that either consists of true aquatic or swamp plants or of terrestrial vegetation that extends from the banks into the water. It is remarkable how robust these fish are and which unfavourable environmental conditions they endure. We caught *Cichlasoma dimerus* in July and August, i.e. at the time of extremely low water levels, in isolated pools that were excessively muddy. Due to vast amounts of rotting plant material these often stank acrimoniously. Besides Catfishes of the genus *Hoplosternum* they were the only fish surviving in this type of water.

Examinations of round about a dozen different localities in Argentina and Bolivia revealed a maximum total hardness of 2°dH, a maximum carbonate hardness of 5°dH, a conductivity of 10 to 90 μS/cm, and a pH between 5.5 and 6.0. The water-temperatures of various collecting sites in the Argentine province of Corrientes were as low as 6.5 to 13°C during July and August.

Care

Although *C. dimerus* may grow up to a length of about 15 cm, its housing does not necessarily require overly large aquaria. A compatible pair can be properly kept in a tank of 125 litres. The bottom should be covered with coarse sand that enables the fish to excavate pits to deposit their larvae during periods of reproduction. Hiding-places between pieces of bog-oak and larger, robust aquatic plants *(Echinodorus, Cryptocoryne aponogetifolia, Vallisneria* sp.) arranged in groups, obviously favour the fish. A fist-sized stone with a smooth surface that can serve as spawning medium completes the decoration.

As can be discerned from the above ecological data, this species is fairly undemanding as far as the water quality is concerned. The tolerance range of 6 to 30°C, with an optimum at 25°C, is an impressive indicator just for how robust this species really is. *C. dimerus* is a voracious feeder that takes the commonly used types of food as they come.

Breeding

this undemanding Cichlid in captivity is easy. *Cichlasoma dimerus* is a typical open-brooder which prefers an exposed flat rock with a smooth surface to spawn on. The beige coloured, well-camouflaged eggs are alternately attended to by both the male and the female. At a water-temperature around 26°C the larvae hatch during the course of the third day from spawning. The parents then immediately transfer them to a small pit they have dug out in advance. Another five days pass until the fry swim freely during which time they are recurrently transferred to several other pits. The school of young fish is eventually guided and protected by both parents. Initial feedings should consist of freshly hatched nauplii of the Brine shrimp. Provided with sufficient amounts of food they grow rapidly and may measure almost ten centimetres after only six months.

Table 21:

Site:	Pond in the Rio Miriñay, ca. 80 km southeast of the town of Mercedes (Prov. Corrientes, Argentina)
Clarity:	clear
Colour:	uncoloured
pH:	7.0
Total hardness:	< 1°dH
Carbonate hardness:	3°dH
Conductivity:	150 μS/cm
Depth:	> 1 m
Current:	none
Water temperature:	9.5°C
Air temperature:	18.0°C
Date:	16.7.1993, 14.30 h

The Genus Copora
FERNANDEZ-YEPEZ, 1969

For decades, the genus *Copora* was considered a synonym of the genera *Cichlasoma* or *Theraps* respectively. When KULLANDER eventually revised and rearranged the formerly heterogenous genus *Cichlasoma* in 1983, he already pointed out that *Copora* would be available for a group of Central American Cichlids, but he did not implement this name in its new function. This step was only taken by ALLGAYER in 1989, on occasion of the description of two newly discovered species of *Theraps* from Mexico. He explicitly paid credit to the generic definition as originally lined out by FERNANDEZ-YEPEZ.

Copora is presently regarded as monotypical which means it contains only a single species. It is characterized by particularly long pectoral fins, a small, low-set mouth with strong teeth, and an elongate body that drastically tapers posteriorly. The special position of this species in the systematic

Localities of *Copora nicaraguensis*

arrangement of the Cichlids is, however, less based on its anatomical and morphological peculiarities, but rather on its reproductive biology. *Copora nicaraguensis* is the only American Cichlid whose eggs are not adhesive and therefore do not attach to the spawning medium.

The forested banks of the Rio Negro are flooded by highwater.

91

◆ Copora nicaraguensis
(GÜNTHER, 1864)

This Central American Cichlid from Nicaragua and Costa Rica was originally described as *Heros nicaraguensis*. Synonyms of this species include *Heros balteatum* GILL & BRANSFORD, 1877, and *Cichlasoma spilotum* MEEK, 1912. Live fish were first imported into Europe in the late seventies. Males can reach about 20 centimetres in length with the females being considerably smaller.

The most indicative black components of the colour-pattern are such as a small caudal spot, a large lateral blotch below the lateral line that stretches over an area equal to the distance from the twelfth to the fifteenth spine of the dorsal fin, and a lateral stripe between the gill-cover and the caudal

peduncle. The intensity of these markings depends on the mood.

Male and female specimens are readily distinguished in this dimorphous and dichromatic species as the sexes differ in size and colouration. Females have shiny, intensely yellow or blueish green colour-zones in the dorsal fin, on the back, and particularly on the lower parts of the head. Their bellies are bright yellowish orange to orange red and the rest of the body has a coppery to golden yellow colouration. Ventral and anal fins are blueish.

The males are less colourful. In fact, the shades of green are less intense and the orange red ventral colouration is generally absent. Some specimens may display a bright red spot immediately behind the base of the pectoral fins though. Most male specimens will display a greenish golden to brass or coppery background colour. Even

Female *Copora nicaraguensis*

semi-adult males are easily recognized by a conspicuous spotted pattern consisting of light and dark dots in all their unpairy fins which is a trait absent in females. Older males develop an enlarged forehead and nape region until the profile outline above the mouth is almost vertical and the head looks rather out of shape and proportion.

Several geographical colour morphs of *Copora nicaraguensis* are known (LOPEZ 1974). Specimens from Lake Nicaragua lack green colours. This form has a purple coloured back, the belly regions of females are yellowish red, whitish in the case of males. The upper head region of the males is bright purple, the lower portion yellowish red to golden yellow. The females have a bright red gular region. This colour morph often has five to six spots on the posterior portion of the flanks that may be vertically enlarged and then form bands.

The most significant

specific trait

that helps to identify this Cichlid is its colour-pattern, in particular the lateral spot and stripe. Other Indicators are up to 20 spines in the dorsal and up to 9 spines in the anal fin (LOPEZ 1974) which are unusually high counts. It is therefore unlikely to confuse this Cichlid with any other species.

Its

natural habitats

are confined to the Atlantic slope of Central America. The localities lie in Nicaragua, especially in the large rift lakes Nicaragua, Managua, and Jiloa, and in northern Costa Rica (drainage of Rio Sapoa and Rio San Carlos). The inhabited waters have a pH between 7 and 9 and are partly of moderate hardness. The fish has been recorded from depths below 15 metres in the large lakes, but it usually resides in more shallow areas. Unattached individuals form large schools

Male *Copora nicaraguensis*

which roam wide areas in search of food. Since the members of those schools are all of the same size, they are probably siblings. As was shown by analyses of stomach contents (comp. THORSON 1976), the fish mainly feed on algae in their natural environment, with mosquito larvae (Chironomidae) and molluscs also being important components in their diet.

The

care

of this recommendable Cichlid does not cause specific problems. In the end, it will, however, require a tank with a length in excess of one metre. Hiding-places add to the well-being of this fish. If plants are to be included in the decoration, hard, sturdy species should be given preference. The bottom should be covered with sand.

Breeding

does not require special preparation. It should be borne in mind that this fish does not attach its eggs to any medium, but that these are laid in a pit. It is therefore imperative to use sand instead of coarse gravel. The brood is cared for in a father-mother-family with a distinct task allotment. At temperatures around 27 °C, the larval development is completed after about 12 days. Initial feedings should consist of larvae of the Brine shrimp and powdered dry food.

The Genus Crenicichla
HECKEL, 1840

Probably consisting of more than ninety species of which only some seventy have to date been dealt with taxonomically (KULLANDER 1994), *Crenicichla* represents the most diverse genus of South American Cichlids next to *Apistogramma*. As the ever increasing number of species has been creating intricacy it has been attempted to group similar forms of this genus in various species-groups that delineate closer phylogenetic relationships.

This approach was instigated by KULLANDER (1981) who defined the *Crenicichla lacustris*-group followed by a differentiation of the *Crenicichla saxatilis*- and *lepidota*-groups in the subsequent year. By now, the *Crenicichla lepidota* species-group concept has, however, turned out to be not monophyletic but heterogeneous (KULLANDER 1986: 83) and has therefore been abandoned. PLOEG (1991) then introduced the *Crenicichla lugubris*, *reticulata*, and *wallacii* species-groups, and LUCENA & KULLANDER (1992) eventually defined the *Crenicichla scottii*- and *Crenicichla missioneira*-groups.

The species of *Crenicichla*, which are also referred to as Pike Cichlids, and the representatives of the genus *Teleocichla* together form the tribe of the Crenicichlines, the latter being so closely related to the former that PLOEG included them in his *Crenicichla wallacii*-group. Some species of *Crenicichla* were temporarily kept in the genus *Batrachops*.

Due to their very typical, unusually elongate, slender bodies, often with an oval, almost roundish cross-section, the members of the genus *Crenicichla* can hardly be confused with any other South American Cichlid. They have a more or less protruding lower jaw and meaty lips. The eyes are situated on the upper half of the head. The caudal fin is round. Most *Crenicichla*-

Localities of *Crenicichla acutirostris*

species are typical large Cichlids with a predatory ecology.

◆ Crenicichla acutirostris
GÜNTHER, 1862

Crenicichla acutirostris is a typical large Cichlid with a total length in excess of thirty centimetres. Single live specimens reached Germany for the first time during the mid and late eighties. Specific traits that help to identify this species include a very slender body shape and an unusually pointed head with a clearly protruding lower jaw.

This fish has a greyish green background colouration which contrasts nicely with an intense loam-yellow back. A pattern of reddish orange spots may be present in the transitory area between these two colourzones. The ventral region is rosy to reddish violet. The lower portion of the head is as bright golden yellow as the iris. The dorsal fin is mainly pale blue, wine-red in its posterior part, and has a narrow signal-red margin. The anal fin and the lower half of the caudal are also wine-red. Depending on the mood there may be a dark lateral stripe

Crenicichla acutirostris

between the angle of the mouth and the base of the caudal fin.

A particularly characteristic trait in the colour-pattern of this Cichlid is the presence of nine to eleven, broad, pitch black lateral bands restricted to the back in an area between the base of the dorsal fin and the upper margin of the lateral stripe. This creates the impression as if the fish had about nine, more or less quadrangular, yellow spots on the upper half of the body.

External sexual characters are not clearly recognizable in this species. Females ready to spawn are, however, easily noted as they have a much increased girth.

Natural habitats

The distribution of *Crenicichla acutirostris* lies in northern Brazil and covers the drainages of a few southern tributaries to the lower Amazon River. Locality records exist from the river systems of Rio Xingú, Rio Tapajós, Rio Maués, and Rio Madeira (Rio Aripuana). The first mentioned water-

bodies are typical clear water rivers that carry a minimum of dissolved minerals (total and carbonate hardness around 1 °dH, conductivity hardly ever exceeding 100 µS/cm) with a pH ranging around 6.5.

Care

This Pike Cichlid obviously requires a very spacious tank, even more so as single specimens show a high degree of intraspecific aggression and viciously attack conspecifics. The decoration of the aquarium should therefore include larger stone flakes and pieces of bog-oak arranged to form cave-like hiding-places. The fish leave plants alone, but dig out pits during the breeding season. It is therefore advisable to limit the vegetation to a few robust single plants.

It must be borne in mind that *Crenicichla acutirostris* is a predatory fish with a tendency towards piscivory. The fish therefore requires rich food that should also include meat of fish and shrimp.

▶ *Crenicichla marmorata*
PELLEGRIN, 1904

This species certainly ranks among the most colourful Pike Cichlids and is highly variable. Juveniles are difficult to distinguish from young *Crenicichla strigata* as both have a spotted head pattern and a broad lateral stripe· with another narrow one right above it. Adult specimens then show rosy to red spots and zones in the centre of the body that may extend onto the ventral side and are bordered boldly with deep black. They may partly resemble·bands or fuse to form a lateral stripe. In contrast to *Crenicichla lenticulata*, the reddish zones do not extend as far as to the base of the dorsal fin.

Localities of *Crenicichla marmorata*

Natural habitats

Crenicichla marmorata is widely distributed in the river system of the lower Amazon River and known from the northern tributaries as well as from the southern ones. Records exist from the drainages of the Rio Madeira, Rio Maués, Rio Curuá-Una, Rio Xingú, Rio Trombetas, Rio Puraquequara, and Rio Jari in northeastern Brazil.

Crenicichla marmorata

▶ *Crenicichla sedentaria*
KULLANDER, 1986

Localities of *Crenicichla sedentaria*

The specimens illustrated here were observed and caught by us in June 1983. They were subsequently examined by KULLANDER who found they were representing an undescribed species that he then named in his monograph on the Cichlids of Peru published in 1986. Observations in the natural habitat indicated that this species may grow longer than 25 centimetres.

Semi-adult specimens show a yellowish brown back and a more whitish belly region. A dark lateral stripe runs at eye-level from the upper lip to the base of the tail, often proceeding right to the hind edge of the caudal fin. The area of the flanks between the rear edge of the gill-cover and the base of the caudal fin is covered with about twelve dark bands that are partly forked or appear in the form of double bands. The upper half of the caudal peduncle is marked with an ocellate spot with a golden yellow aura. The caudal carries four or five vertical lines. The dorsal fin is coloured in shades of pale brownish red and has a narrow orange red margin. The shades of red vanish in mature specimens and their unpairy fins assume a yellowish green colouration. The iris is blood-red, and the lateral stripe disintegrates into about twelve irregular blotches. Male and female specimens are isomorphic and therefore difficult to distinguish. Old males have

Juvenile *Crenicichla sedentaria* from Rio Chinipo

Crenicichla sedentaria (Rio Chinipo) in the natural habitat caring for its offspring

larger fins, and their anal and dorsal fins are more pointed and longer.

Natural habitats

The fish occurs in the Rio Napo and Putomayo in Ecuador as well as in the Huallaga and upper Ucayali in Peru. We took photographs and caught specimens in the small clear water river Rio Chinipo. A water analysis on site revealed a low hardness and a clearly alkaline pH. The river bed was predominantly rocky as there was a considerable current. Underwater observations showed that adult specimens even stayed in sections with a strong current. Breeding specimens had, however, chosen bays and backwaters with a slow current. Semi-adult fish were preferably seen in the extreme shallow areas of water remains where they were hiding between dead leaves and branch debris ambushing for prey. Other fishes found in this biotope were the Cichlids *Cichlasoma amazonarum*, *Bujurquina robusta*, and *Bujurquina labiosa,* Sucker Catfishes of the genus *Chaetostoma*, Characins representing the genera *Phenacogaster*, *Brycoamericus, Hemigrammus, Characidium,* and *Astyanax,* and a Freshwater Ray.

Care

The keeping of this fish should take into consideration the observations made in the natural biotopes that are also applicable to many other species of *Crenicichla*. Due to the size of these Cichlids, their long-term care will only be possible in a tank whose length considerably exceeds one metre. Cave-like hiding-places between larger rocks and roots significantly contribute to the well-being of the fish and also are a preconditions for breeding attempts. The fish are fairly robust and antagonistic so that they can only be kept in the company of Cichlids with a similar temperament. In nature, adult fish probably feed mainly on other fish. Their captive diet should therefore include regular offerings of cut pieces of meat of fish, crabs, and shrimp.

As the home waters of this species have no extreme chemical properties,

breeding

should not be a problem. Observations in the natural biotopes revealed that this Cichlid spawns among rocks and roots in cave-like hiding-places much in the same fashion as other *Crenicichla*. Both parental fish participate in caring for their young. As the underwater photograph depicts, they always stay near the school of young fish to guide and protect it.

Table 22:

Site: Rio Chinipo (clear water river) ca. 10 km south of Chicosa (Depto. Junin, Peru)	
Clarity:	3 m
Colour:	greenish
pH:	7.9
Total hardness:	4 °dH
Carbonate hardness:	4 °dH
Conductivity:	116 µS at 26 °C
Depth:	up to 2 m
Current:	swift
Water temperature:	26 °C
Air temperature:	30 °C
Date:	23. 6. 1983, 11.00 h

The Genus Geophagus
HECKEL, 1840

The year 1976 saw the publication of the revision of this genus by GOSSE who recognized ten species. Shortly thereafter it was again necessary to revise the species of *Geophagus* as their systematic arrangement was not monophyletic, but still heterogeneous. KULLANDER (1986) separated a number of species which were partly transferred to the genus *Satanoperca,* but partly refrained from assigning others to specific genera. These are therefore referred to here as *"Geophagus",* i.e. with the generic name in inverted commas.

At present, the genus *Geophagus* is comprised of eleven taxonomically revised species which are complemented by at least ten undescribed forms. These some twenty Cichlids differ from other species of the tribe of the Geophagines in their peculiar morphology of the swim-bladder and spinal cord.

It is particularly noteworthy that there are various reproduction strategies and parental care patterns present among the species of *Geophagus.* As far as is known to date the majority of species appear to be mouthbrooders. Some of them represent the ovophile type, while others are larvophile and do not keep the eggs in the mouth, but only the larvae. *Geophagus argyrostictus,* in contrast, is a typical open-brooder.

Floating houses are the Amazonian response to the extreme seasonal changes in the water level.

◗ *"Geophagus" brasiliensis*
(QUOY & GAIMARD, 1824)

This species ranks among the longest known Cichlids and was described as a species of *Chromis* as early as more than 150 years ago. There is quite a number of alleged synonyms that include *Chromys unipunctata*, *C. unimaculata*, and *C. obscura* CASTELNAU, 1855, *Acara gymnopoma* GÜNTHER, 1862, and *A. minuta* HENSEL, 1870. Some of these names, however, certainly refer to good species. According to today's knowledge, *"Geophagus" brasiliensis* is merely a collective name urgently requiring a revision (LUCENA & KULLANDER 1992).

This Cichlid may reach almost 30 cm in length. It is remarkable for its marvellous shiny, greenish golden scales and reflecting blueish green spots and dashes on its fins. Approximately in the centre of the body and immediately below the upper lateral line it has a black lateral blotch. The iris has a golden colouration. A sexual dimorphism that would allow a clear distinction between

Localities of *"Geophagus" brasiliensis*

the sexes is largely absent. Only old males develop a bulkier head and more elongated dorsal and anal fins.

Specific traits

permitting an identification include the low number of only 8 to 12 gill-rakers situated

"Geophagus" brasiliensis

on the lower portion of the first gill-arch, and 15 to 16 scales around the caudal peduncle. For aquaristic purposes the colour features are more relevant, e.g. the fact that adult specimens do not possess any black markings except the lateral blotch. As far as colours are concerned,

similar species

are rather found among the species of *"Aequidens"* than in the genus *Geophagus*. They, however, have a differently shaped mouth and profile. A similar colour-pattern is for example found in the Silverseam-cichlid (*"Aequidens"* sp.), but this species is characterized by its white margin of the caudal fin. The

natural habitats

of *"Geophagus" brasiliensis* lie in the southeast of Brazil. In comparison with other South American Cichlids, its distribution is rather small with localities confined to regions near the coast between the town of Salvador and the border area to Uruguay. The

care

of this Cichlid in captivity does usually not entail problems since it is hardy and can adapt to a number of different environments. The chemical properties of the water do therefore not need close monitoring. On the other hand, its successful long-term care requires a spacious tank of at least one metre in length as this fish grows to considerable size. The bottom should be covered with sand or very fine-grain gravel as this representative of this genus like others has the habit of searching for food in the ground. The decoration of the aquarium should include hiding-places between pieces of bog-oak and rock flakes as these provide a feeling of safety for the fish. Vegetation must be limited to large single specimens of plant since more delicate ones are

This Indian girl was a skilled aide catching fish in Peru.

usually damaged. *"Geophagus" brasiliensis* accepts all commonly used types of food and thus is easy to cater for. From reaching a length of about twelve centimetres they require a rich diet and should therefore be supplied with pieces of fish and shrimp meat. This is of particular importance when specimens are to be conditioned for

breeding.

A compatible pair that has preferably bonded itself within a group of young fish will spawn willingly. *"Geophagus" brasiliensis* is an openbrooder which rears its fry in a parental family. Once the young fish swim freely they can be fed with newly hatched larvae of the Brine shrimp.

◗ *"Geophagus" pellegrini*
REGAN, 1912

Together with *"Geophagus" crassilabris* STEINDACHNER, 1887, and *"Geophagus" steindachneri* EIGENMANN & HILDEBRAND, 1910, this species forms a small group of closely related Cichlids which have presently to be dealt with as so-called *"Geophagus"*. Although they have been separated from *Geophagus* in a restricted sense by KULLANDER (1986), they have not yet been assigned to another genus.

Adult specimens of *"Geophagus" pellegrini* have a yellowish to yellowish brown background colouration. Their flanks have pale greenish reflections, and particularly the fins of males are orange red. The caudal peduncle is marked with a large black blotch. Other black elements of the lateral pattern largely depend on the prevailing mood and may consist of about five broad bands. The third one of these bands lies

Localities of *"Geophagus" pellegrini*

approximately below the posteriormost spines of the dorsal fin and above the anteriormost one of the anal fin. It is continued on the dorsal fin in the form of a dark spot.

Old males of this species are characterized by a large, hump-like, red bulge.

Male of *"Geophagus" pellegrini*

Their thick lips are conspicuously whitish to bright yellow in colour. Male specimens may grow to a total length in excess of twenty centimetres while females are considerably smaller.

Natural habitats

"Geophagus" pellegrini is endemic to western Colombia. The specimens serving as type series for its description were caught in the vicinity of Tadó, a village in the drainage of the Rio San Juán. More localities are for example known from the river systems of the Rio Atrato and Rio Baudó.

Ingemann HANSEN (in litt.) was kind enough to supply me with the results of his examination of a locality of *"Geophagus" pellegrini* undertaken six or seven kilometres south of the town of Istmia (province Choco) in the drainage of the Rio San Juán. According to this, the Cichlid there lives in very soft water with slightly acidic properties. Carbonate and total hardness were below the detectability-level at this site. The pH-value was 6.7, the water-temperature 24.2 °C, and that of the air 29.8 °C.

Care

Due to the expected adult size, these fish should not be kept in aquaria that fall short of one and a half metres in length. They are to be decorated with sand, pieces of bog-oak, and larger rocks that are arranged to create hiding-places. Since the fish chew through the bottom substrate, coarse gravel would obviously be the worst choice. This species is very plant-tolerant. If aquatic vegetation is used, attention must be given to the fact that the plants do not constrain the available swimming space and that their roots are covered with large stones so that the fish cannot dig them out when they chew through the substrate.

Co-inhabitants of the same tank should not be aggressive species, but preferably those Cichlids that also belong to the tribe Geophagini.

Adult specimens of *"Geophagus" pellegrini* require rich food and it is advisable to include regular offerings of fish and shrimp meat in their diet. This Cichlid has proven to be a robust, adaptable aquarium fish that does not pose particular demands on the chemical qualities of the water.

Breeding

An attempt to breed should begin with the transfer of one male and several females into a separate spacious breeding tank which offers a number of hiding-places. *"Geophagus" pellegrini*, like the other two species mentioned, belongs to the ovophile mouthbrooders and is thus a great exception among South American Cichlids whose majority is comprised of either open-brooders or larvophile mouthbrooders.

It is typical of this Cichlid that there is no lasting bond between the reproductive partners, i.e. males and females come only together for the purpose of spawning. Following a vivacious courtship, the female expels small numbers of two to four eggs onto a solid surface, preferably a root or a stone. She then immediately takes them into the mouth and retreats to a covered spot where she awaits the things that are due to happen.

In contrast to other South American Cichlids, the male does not participate in the parental care for the offspring. Depending on the water-temperature it takes about fourteen days until the larval stage is completed and the young fish are released from the mouth of the mother fish for the first time. At this point of time they are already large enough to cope with newly hatched Brine shrimp so that their rearing is easy.

▶ "Geophagus" steindachneri
EIGENMANN & HILDEBRAND, 1910

The real identity of this Cichlid was for a long time subject to controversial discussions as the synonyms G. *hondae* REGAN, 1912, and G. *magdalenae* BRIND, 1943, were available and there was uncertainty which one of the names would be valid according to the rules of nomenclature. This problem was eventually solved by GOSSE & KULLANDER in 1981.

The maximum body length of males can clearly exceed 15 centimetres although most of the specimens are distinctly smaller. On a grey or yellowish background colouration, particularly the scales on the posterior half of the body are marked with shiny green dots. The flanks often have an irregular pattern of bands and stripes. The most conspicuous feature of adult males is a well-developed bulge on the forehead and nape

Localities of *"Geophagus" steindachneri*

that is coloured wine-red. Its size largely depends on whether or not there are other males in the same tank and on the rank a fish has within this group. Female specimens are much smaller and lack such humps.

Male *"Geophagus" steindachneri*

The most significant

specific trait

that identifies this Cichlid, is the bulged forehead as delineated above. This species differs from the two other forms of this group in having no spotted or banded pattern on the flanks. The

natural habitats

of *"Geophagus" steindachneri* are restricted to the northeast of South America within the borders of Colombia. Localities known to date include the drainages of the rivers Magdalena, Sinú, Canda, and Limón. As long as the specimens are not fully grown, their

care

necessitates an aquarium of only about one metre in length. It should be furnished with sand or very fine gravel, larger rocks, and pieces of bog-oak. A few thickets of plants which partition the ground-space and simplify the constitution of territories may be used in addition. The chemical properties of the water need no close monitoring as *"Geophagus" steindachneri* is robust and adapts to various environments. It can be kept in company of other Geophagines or representatives of the genus *Aequidens* of approximately its own size, if the size of the aquarium permits.

Breeding

this Cichlid is not difficult as it readily spawns under the most different keeping conditions. Amongst the South American Cichlids it assumes a special position in so far as it is one of the very few specialized ovophile mouthbrooders which rear their fry in a mother-family. This form of parental care is otherwise common in East African Cichlids. It entails that there is no bonding between male and female. The reproduction partners come together for the sole purpose of reproduction and separate

White water biotope in the central Ucayali near Pucallpa in Peru

immediately thereafter. It is therefore sensible to keep a male together with a number of females to prevent that only one female fish is constantly exposed to the harassment of the male which is more or less sexually active all the time. Some behavioural patterns indicate that the brownish red spot at the angle of the mouth of the males serves intraspecific communication purposes and is the key trigger for the female to release eggs.

◗ *Geophagus surinamensis*
(BLOCH, 1791)

This is one of the Cichlids longest known to science. In its original description it was allocated to the genus *Sparus* which does, however, not belong to the family Cichlidae.

In his 1976 revision of the genus *Geophagus*, GOSSE still favoured the opinion that *Geophagus surinamensis* was a polymorphous species with a vast distribution. KULLANDER & NIJSSEN (1989) then demonstrated that this Cichlid only occurs in a very confined area in the Guyanas. A consequence of this is that the various reports under this name in the aquaristic literature up to the late eighties probably all refer to other species, for live specimens of this fish have first been imported in the early nineties.

Geophagus surinamensis is one of the less conspicuously coloured representatives of its genus. The pattern of about five alternat-

Localities of *Geophagus surinamensis*

ing, narrow, yellow and green glossy bands that cross the flanks vertically, so common in other species, is comparatively weakly developed in this species. There is a yellow colour-zone that stretches from the base of the pectoral fins to that of the ventral fins. The ventral fins and all unpairy fins are basi-

Geophagus surinamensis, wild-caught specimen from Guyana

cally pale wine-red. The caudal fin has a pattern of fairly large, roundish spots and similar markings are also found in the soft sections of the anal and dorsal fins.

The black pattern is reduced in this species to a small lateral spot and five or six lateral bands that may under the influence of certain stages of excitement appear as indications on the flanks between the edge of the gill-cover and the rear margin of the dorsal and anal fins. The adult size of this species may be fifteen centimetres.

Specific traits

The genus *Geophagus* contains a number of species that resemble each other very closely so that the determination of live fish is often fairly difficult. Features in the colour-pattern that allow an identification of *Geophagus surinamensis* include the absolute lack of black markings on the head and the pattern of large, roundish, pale blue spots that covers the entire caudal fin. This is complemented by a relatively small lateral spot that generally covers a vertically expanded area of three scale rows from between the eleventh to the thirteenth scale of the lateral line. A dark spot below the posterior edge of the dorsal fin is absent and in comparison with related species the pattern of horizontal stripes on the flanks is weakly developed.

Natural habitats

In contrast to what has been indicated in the scientific literature earlier, *Geophagus surinamensis* in fact has a very small distribution. It only covers the east of Surinam and the extreme west of French Guyana. Localities are mainly known from the river systems of the Saramacca, Surinam, Marowijne, and Maroni (KULLANDER & NIJSSEN 1989).

Care

In the case of *Geophagus surinamensis* and other "Eartheaters", it must be ensured that the bottom of the aquarium is covered with a thick enough layer of fine sand that enables the fish to chew through the substrate. This type of searching for food is instinctive and an adaption to sandy biotopes. Keeping this species in a tank with coarse gravel would therefore be absolutely inappropriate.

To feel comfortable, the fish furthermore require some hiding-places among rock flakes or pieces of bog-oak. Larger plants are suitable decoration items as this Cichlid is said to be plant-tolerant.

Breeding

The reproductive biology of the species of *Geophagus* is of particular interest in so far as observations in captivity have revealed (STAWIKOWSKI & WERNER 1988, a.o.) that there are a number of different strategies in this genus. These include larvophile and ovophile mouthbrooders as well as openbrooders. Although different patterns of reproduction and parental care are noteworthy with regard to the evolution of these behavioural strategies, they are not too unusual for South American Cichlids. This situation can also be found in the genera *Gymnogeophagus*, *Aequidens*, and *Heros*.

To date, little precise is known about the parental care behaviour of *Geophagus surinamensis*. This is only superficially in opposition to the numerous breeding reports that have been published in the aquarium literature. According to today's knowledge these all referred in fact to other species of this cladistic assembly.

Breeding attempts should take into consideration that most water-bodies in the Guyanas have very soft and acidic water.

The Genus Guianacara
KULLANDER & NIJSSEN, 1989

Guianacara is one of the genera of Cichlids that have been defined only recently. At present, it accommodates four species, but there is at least another one which is as yet undescribed. As the name already suggests, the distribution of this systematic unit centres in the three Guyanas. Moreover, there are localities known from the river systems of the Rio Caroni, the Rio Branco, and the Rio Trombetas in the adjacent regions of Venezuela and Brazil.

The species of *Guianacara* are closely related to those of the genus *Acarichthys*, but also show an affiliation to the genera *Biotodoma*, *Geophagus*, and *Satanoperca*.

All *Guianacara*-species share the laterally greatly compressed, deep body-shape which has its greatest depth between the base of the ventral fins and the anteriormost rays of the dorsal fin. Particularly in old males, the head also appears extremely deep and short. The small level mouth lies relatively low in the skull. These fishes are not particularly colourful. Generic traits are such as black super- and suborbital stripes and a broad band over the centre of the flanks that may be reduced to a lateral blotch and is of value to distinguish species. A feature shared by juveniles of all species is the deep black colouration of the membranes between the first four spines of the dorsal fin.

Localities of *Guianacara geayi*

▶ *Guianacara geayi*
(PELLEGRIN, 1902)

This Cichlid was originally described as *Acara geayi* and subsequently kept as a member of the genus *Aequidens* for decades. KULLANDER (1980) then had temporarily assigned it to the genus *Acarichthys*.

Guianacara geayi is no conspicuously coloured fish. The flanks are blueish grey to greyish brown and show a pattern of small, glossy, greenish golden speckles in about nine horizontal rows. Old males in particular also have such glossy spots on their caudal, anal, and dorsal fins. These are large in comparison and, in the case of the dorsal, even extend onto the spinalous area.

The contrasting black markings are most conspicuous although their intensity varies with the prevailing mood. They consist of an ocular band that runs from the nape to the lower edge of the gill-cover, and either a broad band over the central flanks that tapers on the belly, or a round lateral blotch situated on the upper portion of the body, but below the lateral line. Juveniles typically have black membranes between the first three spines of the dorsal fin.

Guianacara geayi female

108

Under aquarium conditions, males can reach approximately twelve centimetres in length with the females being much smaller. A distinct sexual dimorphism only becomes visible with age. Besides the difference in size the more produced fins of the males and the more sharply defined spotted pattern of their unpairy fins then provide useful hints on the sexual identity of specimens.

Specific traits

in the colour-pattern of *Guianacara geayi* that differentiate it from other representatives of this genus include the broad, evenly intense band that tapers towards the belly and the glossy spots that are fairly large in comparison with other species and are also present on the spinous portion of the dorsal fin. It is furthermore of importance that adult specimens lack a black colouration of the membranes between the anterior spines

of the dorsal fin and that the lateral spot lies below the lateral line (KULLANDER & NIJSSEN 1989).

Similar species

Adult *Guianacara owroewefi* are characterized by having black anterior membranes in the dorsal fin, in *G. sphenozona* it is the lateral band that is often reduced to a spot on the back, and *G. oelemariensis* has no elongated fin membranes in the anterior section of the dorsal fin and the lateral spot, replacing the lateral band, lies in the centre of the flank. Its yellow coloured gill-cover is typical of *Guianacara* sp. from Venezuela.

Natural habitats

The type specimens of this species originate from the drainage of the Camopi River in French Guyana. According to our present knowledge the distribution of *Guianacara*

Guianacara geayi, male

geayi is restricted to that country with further confirmed localities being known only from the river systems of the Oyapock and the Approuague (KULLANDER & NIJSSEN 1989).

Care

An adequate husbandry of this Cichlid requires an aquarium of about one metre in length. It should be furnished with sand and hiding-places among pieces of bog-oak and rocks. This fish is undemanding as far as food and water-values are concerned.

Breeding

This Cichlid will also readily breed in moderately hard water with a slightly alkaline pH. It should be taken into consideration that this fish is a cavebrooder and prefers to spawn on a vertical cave wall. Parental care takes place in a father-mother-family with both sexes assuming different tasks. The female mainly cares for the brood while the male keeps potential predators out of the breeding territory.

Localities of *Guianacara* sp.

▶ *Guianacara* sp.
(Rio Caroni)

This representative of the genus *Guianacara* from Venezuela is yet to be scientifically described. It has a loam-yellow to greyish yellow background colouration. The poste-

Guianacara sp. (Rio Caroni) from Rio Cucurital

rior portion of the gill-cover has an orange yellow shade. In certain light situations the flanks show eight or nine horizontal rows of the small glossy golden spots. The ventral and anal fins as well as the hind part of the dorsal fin have a pale orange yellow tinge.

The black markings of adult fish are limited to a round lateral spot below the lateral line and an ocular band that extends from the nape down to the lower angle of the gill-cover. The bands of either side are connected with one another by a soot-coloured brachiostegal membrane. In contrast to adult specimens juveniles up to a length of about three centimetres have a broad lateral band rather than a lateral spot. They also display a deep black colouration of the membranes between the first three spines of the dorsal fin.

In captivity, male fish can grow up to nearly twelve centimetres in length while females are considerably smaller. Other than that there are almost no differences between the sexes except that fully grown males have larger fins.

Natural habitats

Locality records of this Cichlid are presently limited to the drainage of the Rio Caroni in Venezuela. We caught it in the Rio Cucurital, a tributary to the central Caroni. In April, at the time of low tide, the river had a swift flow and the rocky river bed was characterized by its numerous rapids. The water was clear, very soft, and acidic. Large numbers of about one centimetre long juveniles were observed in calmer sections along the banks where they found cover among boulders and cobble. At this site, we simultaneously recorded several undetermined Characins and a species of *Crenicichla*.

Care

This species of *Guianacara* is one of the more robust and adaptable fish for the aquarium. A compatible pair will be suit-ably housed in a tank of about one metre in length. It should be decorated with sand and a few larger rocks and stone-flakes that are arranged to create several caves. Additional decoration items may be roots and some large specimens of aquatic plants. The fish are undemanding regarding food and chemical properties of the water.

Breeding

An attempt to breed this species should take into consideration the water-values of the natural habitats as otherwise there is a chance that eggs and larvae do not develop properly. This Cichlid prefers to spawn at a well-hidden site and therefore usually chooses a cave for this purpose. In contrast to specialized cavebrooders it, however, does not attach the eggs to the ceiling of the cave, but to the walls.

The brood is being cared for in a father-mother-family in which both parental specimens assume different responsibilities. The female cares for the eggs and larvae inside the breeding cave while the male keeps predators out of the breeding territory. When the juveniles have completed their larval development and begin to form a school, both parents jointly care for and protect them. Rearing the young fish on an initial diet of larvae of the Brine shrimp is easy.

Table 23:

Site: Rio Cucurital (Rio Caroni drainage) in Venezuela (Estado Bolivar)	
Clarity:	> 1 m
Colour:	uncoloured
pH:	6.0
Total hardness:	< 1°dH
Carbonate hardness:	< 1°dH
Conductivity:	< 10 µS/cm
Depth:	> 2 m
Current:	strong
Water temperature:	29°C
Air temperature:	32°C
Date:	6.4.1992, 13.00 h

The Genus *Gymnogeophagus*
RIBEIRO, 1918

The distribution of the species of *Gymnogeophagus* is confined to the eastern parts of central South America which means they do not occur in the Amazon region, but only in the countries Paraguay, Uruguay, Argentina, and the south of Brazil. Locality records of these Cichlids are primarily known from the drainages of the Rio Paraguay, the Rio Paraná, the Rio Uruguay, and a number of smaller rivers on the Atlantic coast.

This systematic unit was last revised by the ichthyologists REIS & MALABARBA in 1988. Today, the genus *Gymnogeophagus* holds eight species, three of which were described only recently. These Cichlids belong to the tribe of the Geophagines and are closely related to the genus *Geophagus*. Their separate position in the systematic organization of the Cichlids is mainly based on peculiarities in the anatomical structure of the skeleton of the dorsal fin. The species of *Gymnogeophagus* differ from all other genera in lacking supraneural bones (particular bones in the skeleton of the dorsal fin). Instead, they have a generic trait in the form of a small, inflexible, forward projected thorn that is not known from any other Cichlid.

It is remarkable how varied the reproductive and parental care strategies are within this genus, i.e. there are open-brooders which spawn on a substrate as well as mouthbrooders. The parental care may involve a family-structure or be just left to the female.

Localities of *Gymnogeophagus balzanii*

The maximum length of this fish should come close to 15 centimetres with the females being considerably smaller. Significant features of the colour-pattern of this species include a black ocular band, a dark lateral spot below the lateral line approximately in the centre of the flank, and nine to ten dark lateral bands that are arranged in pairs, but are often reduced to weak indications.

Sexes are easy to distinguish as there is a distinct sexual dimorphism present. Adult males grow an unshapely, balloon-like enlarged head and long tapering anal and dorsal fins. Furthermore, they have far more and larger green glossy spots on the gill-cover and on the anterior parts of the flanks. Female specimens on the other hand display bright orange yellow colours on the lower head, the gular, and chest region.

◆ *Gymnogeophagus balzanii*
(PERUGIA, 1891)

Synonyms of this taxon include *Geophagus duodecimspinosus* BOULENGER, 1895, and *Gymnogeophagus cyanopterus* RIBEIRO, 1918.

Specific traits

include the pattern of double bands on the flanks, the unique shape of the males' heads, and the comparatively high number of 12 to 15 rays in the dorsal fin.

112

The
natural habitat

of this Cichlid lies in northeastern Argentina, parts of Uruguay and Paraguay, and southeastern Brazil. Localities are known from the drainage systems of the Rio Paraguay, Paraná, Rio Uruguay, and the headwaters of the Guaporé. *Gymnogeophagus balzanii* thus also inhabits water-bodies in the subtropical portion of South America where water-temperatures may drop to below 10 °C in winter.

Care

This Cichlid is fairly easy to cater for as it is adaptable and undemanding with regard to the chemical properties of the water. Long-term, it will require an aquarium with a length in excess of one metre. The tank should be decorated with sand or very fine gravel and with pieces of bog-oak and rocks that are arranged to form hiding-places. As an additional decoration item larger single specimens of robust aquatic plants such as *Echinodorus* may be chosen. The fish are relatively peace-loving and can therefore be kept in company of other Geophagines or species of *Aequidens*. Its diet should entail cut meat of fish and/or shrimp.

A diet rich in proteins is a precondition for

breeding.

This is not particularly difficult and will also be successful in moderately hard to hard water and at an alkaline pH. *Gymnogeophagus balzanii* is a larvophile mouthbrooder which rears its fry in a mother-family. The eggs remain on the spawning medium until the larvae hatch. These are then taken into the mouth by the mother fish for further development. The male does not participate in this care.

Gymnogeophagus balzanii, female

The Genus Herichthys
BAIRD & GIRARD, 1854

For decades, *Herichthys* was considered to be a synonym of *Cichlasoma*. When this heterogenous genus was eventually revised, the name was revalidated and now applies to a group of Central American Cichlids (KULLANDER 1983, ARTIGAS-AZAS 1993). At present it contains about half a dozen species, but as this clade is insufficiently separated from related genera it still requires a revision.

Localities of *Herichthys carpintis*

▶ Herichthys carpintis
(JORDAN & SNYDER, 1899)

The Pearl-cichlid was originally described as *Neetroplus carpintis,* and until very recently it was referred to as *Cichlasoma cyanoguttatum* in the aquarium literature. It may reach a length of more than thirty centimetres.

The entire body of this fish is covered with a dense pattern of blueish green glossy spots that may partly extend onto the fins.

The background colour is blackish to greyish green. Subadult specimens usually show five or six black spots on the posterior half of the body arranged in a horizontal row. These may also be vertically enlarged to form a banded pattern. The markings fade with increasing age.

The Pearl-cichlid has an unusual breeding colouration. The anterior third of the

Herichthys carpintis

body at those times assumes a particularly light tone that however does not affect the lower jaw, the throat, and the chest, and therefore creates a stunning contrast to the very dark colouration of the rest of the body. This light zone has about the shape of a right-angled triangle whose base line extends from the upper lip over the forehead to the nape. Older specimens develop a distinct sexual dimorphism with the males not only becoming much larger, but also growing an unshapely hump on the forehead.

Specific traits

of *Herichthys carpinis* are the dense pattern of relatively large glossy spots and the enormous hump on the forehead of older males.

A similar species

which might cause confusion is first and foremost *Herichthys cyanoguttatus* BAIRD & GIRARD, 1854. This species differs in having much smaller glossy spots and its males lack the distinctive hump on the forehead. The

natural habitats

of *Herichthys carpinis* lie in the border region of the Mexican states of Tamaulipas, Vera Cruz, and San Luis Potosi. Undoubted records come from the Laguna de Carpinteros and the drainage of the Rio Panuco. The type specimens originate from the vicinity of the Town of Tampico. We managed to find large numbers of this Cichlid in the tributaries of the Laguna Media Luna near the town of Rio Verde during spring 1983. Following TAYLOR & MILLER (1983) the upper drainage of the Rio Verde does, however, not belong to the naturally inhabited region as the species was introduced here. The long-term

care

of the Pearl-cichlid can only be recommended if an aquarium of about one and a

Biotope of *Herichthys cyanoguttatus* in northern Mexico (Rio Pablillo, Nuevo León)

half metres in length is available because this fish reaches a considerable size and single individuals show a high degree of intraspecific aggression.

Monitoring of the chemical qualities of the water can largely be neglected as this species is robust and adaptable. The decoration of the tank should include larger rocks that are arranged to form caves and simplify the definition of territorial borders. Delicate plants have little chances for survival. Due to its size, this Cichlid requires a rich diet that can be ensured by regular offerings of fish and shrimp meat.

As specimens mature on reaching a length of about 10 cm, half-grown specimens can already be used for

breeding.

They spawn on a stone in the fashion typical of openbrooders. Both parents care for the eggs and the fry. Initial feedings for the young fish should consist of nauplii of the Brine shrimp and crushed flake-food.

▶ *Herichthys cyanoguttatus*
BAIRD & GIRARD, 1854

Localities of *Herichthys cyanoguttatus*

The Texas-cichlid was originally described as *Herichthys cyanoguttatus*. The fact that it differs from the genus *Cichlasoma* with its exclusively conical teeth by also having flattened, blunt teeth was considered sufficient to accommodate it in its own special genus. This point of view was, however, opposed right from the beginning and never accepted by a number of ichthyologists. Subsequently it was found that the shape of the teeth was largely dependent on the feeding ecology of the fish and their environment. The shape of the teeth of these Cichlids is therefore not genetically fixed, but based on a process of wear. This is the reason why *Herichthys* was temporarily degraded to a section of *Cichlasoma*.

A number of authors regard *Neetroplus carpintis* JORDAN & SNYDER, 1899, *Heros teporatus* FOWLER, 1903, and *Heros pavonaceus* GARMAN, 1881, as synonyms of *Herichthys cyanoguttatus* (comp. TAYLOR & MILLER

1983). Based on our observations made in Mexico we cannot share this opinion and therefore follow ALVAREZ (1970) and LOISELLE (1982) who acknowledge several species or subspecies respectively.

Herichthys cyanoguttatus can reach a length of almost 25 centimetres. The body of this fish is greyish green to dark beige in

Herichthys cyanoguttatus

colour. The caudal peduncle and also the centre of the flanks are each marked with a black spot. The posterior half of the body is usually patterned with four or five, broad, dark bands. The first of these bands sometimes covers the lateral spot. Body and fins are densely marked with greenish silvery to pale green glossy dots that assume a bright green tone on the head. The anterior portion of the body lightens up considerably during periods of parental care. Simultaneously the posterior part of the body as well as the lower portions of chest and throat become almost black creating a sharp contrast. As there is no sexual dimorphism recognizable in young specimens, distinguishing sexes is difficult. Older males then develop larger fins. The most significant

specific trait

that identifies this Cichlid is the colour-pattern. In contrast to related species the body and fins of *Herichthys cyanoguttatus* are covered with very small glossy spots.

Similar species

that might cause confusion include *Herichthys carpintis* (JORDAN & SNYDER, 1899). For decades this fish has erroneously been referred to as *Herichthys cyanoguttatus* in Europe. This species has relatively large blueish green glossy spots though. Moreover, older males of this species grow a conspicuous hump on the forehead that is absent in the Texas-cichlid.

Herichthys cyanoguttatus is the species with northernmost distribution. Its

natural habitats

lie in the Mexican states of Coahuila, Tamaulipas, and Nuevo León extending into the southernmost parts of Texas so that the name Texas-cichlid is not too far-fetched. Localities have been recorded from the lower Rio Nueces, the lower drainage of

the Rio Grande (Pecos and Devil's River), and in the Rio Conchos. The type specimens were collected in Brownsville, Texas. The specimens illustrated here were observed by us in the Rio Pablillo (Nuevo León) which forms part of the upper drainage of the Rio Conchos. This is a relatively clear, fast flowing mountain stream coming down from the Sierra Madre Occidental. The fish preferably resided in a swamp that is crossed by numerous creeks and rivulets rich in aquatic vegetation *(Hydrocotyle verticillata, Ludwigia sp., Lobelia cardinalis)*. In its natural environment this Cichlid is an omnivore with a high percentage of plant matter in its diet.

Care

Although this species may withstand temperatures as low as 15 °C for short periods, ideal keeping conditions should focus on temperatures between 22 and 25 °C. It requires large aquaria furnished with fine gravel, larger rocks, and pieces of bog-oak. Greening the tank is useless as the fish will eat the plants.

Breeding

is not difficult. The fish are openbrooders and care for their young in a father-mother-family with a distinct task-allocation.

Table 24:

Site: Rio Pablillo (Nuovo León, Mexiko), about 15 km west of the town of Linares	
Clarity:	< 2 m
Colour:	uncoloured
pH:	8.95
Total hardness:	17.0 °dH
Carbonate hardness:	11.5 °dH
Conductivity:	6400 µS at 24.5 °C
Depth:	< 150 cm
Current:	partly strong
Water temperature:	24.5 °C
Air temperature:	24.0 °C
Date:	27. 3. 1983, 11.00 h

▶ *Herichthys tamasopoensis*

ARTIGAS-AZAS, 1993

Localities of *Herichthys tamasopoensis*

Although this Cichlid closely resembles *Herichthys cyanoguttatus* BAIRD & GIRARD, 1854, the original description fails to provide a differential diagnosis between these two species. A difference in the colour-pattern appears to lie in the lateral glossy spots though, which are more numerous and smaller in *Herichthys cyanoguttatus*.

The basic colouration of the fish becomes much lighter during periods of parental care. Simultaneously the dark body pattern turns deep black. This is complemented by a black colour-zone on the lower surface that extends from the lower lip to the anal fin widening immediately behind the gill-covers to reach the pectoral fins. This species may reach close to twenty centimetres in length.

Herichthys tamasopoensis appears to be endemic to the drainage of the Rio Gallinas, a tributary to the Rio Panuco. At the waterfalls of Tamasopo, a collecting site of this species, we recorded the following data in March 1983: pH 8.35, GH 8°dH, KH 10°dH, air-temperature 30°C, water-temperature 24°C, conductivity 3850 µS/cm.

Pair of *Herichthys tamasopoensis* involved in parental care in the wild (comp. photo in the front)

The Genus Heros
HECKEL, 1840

This Cichlid genus was at times very large as it used to contain the majority of species of American Cichlids which subsequently had to be re-allocated to the genus *Cichlasoma* due to reasons of nomenclatural priority. As a result of the revision and substantial reduction of the heterogenous genus *Cichlasoma* by KULLANDER (1983), *Heros* again received valid status as a South American Cichlid genus. After a decade of being considered monotypical, more recent research suggests that there are at least two species (KULLANDER 1986, 1994) which are more closely related to the genus *Symphysodon*. A similarly deep, laterally compressed body and comparable particulars in the anatomy of the posterior body and the structure of the swim-bladder are also found in the representatives of the genera *Pterophyllum* and *Mesonauta*.

The genus *Heros* has a wide distribution. Representatives occur in the north, from the drainage of the Orinoco and parts of Guyana, through the Amazon river system, up to the upper Rio Guaporé in the south.

In many American Cichlid biotopes the fish find cover among Water hyacinths.

▶ *Heros appendiculatus*
(CASTELNAU, 1855)

As this Cichlid reaches a length of more than 25 centimetres, it is one of the most favoured fish for food purposes and can therefore be regularly found on the fish markets of its home countries. The fish are often speared by native Indians who stalk them at night with torchlights.

Depending on the place of origin, the colour-pattern is greatly variable. The most significant specific traits include a roundish black spot on the bases of the soft part of the anal fin and the dorsal fin respectively. These two are often connected by a wide black band. In certain moods there may be another six to seven vertical bands that, however, often appear as mere indications. Adult specimens display the banded pattern only on the lower half of the flanks. Furthermore they can be identified by their bright orange yellow ventral and anal fins.

Localities of *Heros appendiculatus*

A species of *Heros* from the lower Rio Negro in Brazil lacks this feature. Instead, it has a very conspicuous pattern of reddish brown spots and vermicular markings on the head and lateral parts of the body. The background colour may be any shade of green to brownish yellow. Through selective

Heros appendiculatus from Rio Ucayali

breeding, aquarists have since produced a partly albinistic variety which displays a bright orange yellow background colour with a pattern of red dots and dashes.

Heros appendiculatus is isomorphic so that there are no reliable external differences between the sexes.

Specific traits

in the colour-pattern of this Cichlid that make identification possible, include the orange yellow coloured ventral and anal fins and the short second to fifth lateral bands that are confined to the lower half of the body in adult specimens and may even be as much reduced as to form elongate spots.

Similar species

The question of how many species of *Heros* there are and how to differentiate them can presently be answered only incomprehensively. In the case of *Heros severus,* lateral bands are also visible on the upper body-half and its sixth band extends less far onto the dorsal fin. Some representatives of the genus differ from others in old specimens having no bands altogether.

Natural habitats

The distribution of *Heros appendiculatus* covers the drainage of the upper Amazon River. Among others, localities are known from Rio Ucayali and the lower sections of the rivers Nanay, Napo, Ampiyacu, Yavarí, Putumayo, and Iça. We found this Cichlid in the drainage of the Ucayali, in water with a slightly alkaline pH. The fish appear to preferably reside in spots covered by floating plants or submerged branchwork.

Taking into consideration the size this Cichlid may eventually attain, its long-term

care

will certainly only be successful in tanks with lengths of one to one and a half metres.

Heros sp. from the lower Rio Negro (Brazil)

To feel at ease, the fish definitely require hiding-places. Plants can only be recommended with reservations as the fish will damage delicate species. Semi-adult, single specimens are often quite aggressive towards conspecifics if the available space is inadequate. Notwithstanding, they can well be kept together with other species. The fish is hardy and adaptable, but no doubt requires a rich diet. Even for

breeding

this Cichlid no close monitoring of the water values is necessary. This is particularly true for specimens from the upper Ucayali. *Heros appendiculatus* is a typical open-brooder that cares for its offspring in a parental family-structure.

Table 25:

Site:	Rio Shahuaya, drainage of the upper Ucayali (Peru)
Clarity:	2 m
Colour:	greenish
pH:	7.5
Total hardness:	10 °dH
Carbonate hardness:	11 °dH
Conductivity:	315 µS at 27 °C
Depth:	< 150 cm
Current:	slight
Water temperature:	27 °C
Air temperature:	28 °C
Date:	14. 7. 1981
Time:	11.00 h

▶ *Heros severus*

HECKEL, 1840

This Cichlid has the deep, laterally compressed body typical of all species of *Heros*. Compared with *Heros appendiculatus* it is slightly more slender though and more elongate (KULLANDER 1986). Juveniles are greenish while adult specimens are rather yellow in their basic colouration. The sides of the body are marked with altogether eight bands whose visibility, however, depends on the prevailing mood. The first of these bands is oblique to connect the nape with the edge of the gill-cover, and the eighth lies on the base of the caudal fin. More often, only the seventh band is clearly recognizable which runs from the posterior rays of dorsal fin to their counterparts in the anal fin, but hardly extends onto the dorsal fin itself. In dominating specimens the iris is conspicuously signal-red. With most scales on the lower half of the body carrying a red dot, adult specimens appear to have a pat-

Localities of *Heros severus*

tern consisting of about eight horizontal rows of spots.

This Cichlid may reach a length of more than 25 centimetres. A distinct sexual dimorphism is absent so that the determination of the sex on the basis of external characters is difficult. Fully grown specimens, however, display a sexual dichromatism in

Heros severus from the Rio Caura drainage

the form of the females carrying no pattern on the lower portion of the head while this area is spotted with reddish brown in the males.

Specific traits

The specific features of the colouration that permit identification of this Cichlid and a separation from other species of *Heros* include first and foremost the pattern of red speckles on the flanks. This trait becomes visible when the fish have grown to a size of about seven centimetres.

Natural habitats

According to the present state of knowledge (comp. KULLANDER 1986), the distribution of *Heros severus* ranges over the drainage systems of the upper Rio Negro and upper and central Orinoco in Venezuela and northern Brazil. The specimen illustrated here was caught by us in 1992, in a tributary to the lower Rio Caura that in turn flows into the central Orinoco.

The collecting site represented a pond-like backwash of a larger stream that was used as a bathing site. This water-body offered ample hiding-places in the form of dense swamp vegetation along its margins. In April, at the time of low tide, it was home to a surprisingly large number of Cichlids. Besides *Heros severus* we also recorded *Satanoperca daemon*, *Mesonauta insignis*, *Acaronia vultuosa*, and also a species of *Aequidens* and another one of *Apistogramma* from here. The clear water was very soft and distinctly acidic.

Care

The adequate captive care of this large Cichlid demands a spacious aquarium. Single specimens are prone to excessive intraspecific fights and can therefore be only kept together if there are enough space and numerous hiding-places available. The tank should be decorated with coarse sand, pieces of bog-oak, and larger stone flakes that make it easy for the individuals to define their territories. Plants can only be recommended with reservations as *Heros severus* considers delicate sprouts an additional source of food. Adult fish require rich food and should regularly be offered small pieces of fish and shrimp meat.

Breeding

Until very recently it was generally accepted that in the tribe Cichlasomini mouthbrooding would be limited to representatives of the two genera *Bujurquina* and *Aequidens*. It therefore came as a sensation when it was discovered that there is also a mouthbrooder among the aquaristically quite well-known species of *Heros* (STAWIKOWSKI 1994). Captive observations have demonstrated that *Heros severus* is a larvophile mouthbrooder. Like substrate-brooders, the fish spawn on a solid surface and care for the eggs in a parental family. When the larvae eventually hatch, the parental fish switch over to a mouthbrooding care.

Breeding attempts should take into consideration the water-values of the natural environment. Otherwise there is a risk that eggs and larvae will not develop normally.

Table 26:

Site: Banjo of Aripao southeast of the town of Maripa (Rio Caura drainage, Venezuela)	
Clarity:	> 1 m
Colour:	uncoloured
pH:	5.3
Total hardness:	< 1°dH
Carbonate hardness:	< 1°dH
Conductivity:	< 10 µS/cm
Depth:	< 80 cm
Current:	slight
Water temperature:	30.0°C
Air temperature:	31.5°C
Date:	10. 4. 1992
Time:	11.00 h

The Genus Herotilapia
PELLEGRIN, 1904

This genus consists of only one species from Central America. Its close relationship to the genus *Archocentrus* is obvious, with the differences lying in the shape of the teeth.

▶ *Herotilapia multispinosa*
(GÜNTHER, 1869)

In its original description, this Cichlid was allocated to the genus *Heros* and later kept as *Cichlasoma* in the literature. Live fish reached Europe for the first time in 1971. In captivity it should reach about 12 centimetres in length.

The colour-pattern of this Cichlid is so variable that it is sometimes referred to as the "Rainbow-cichlid". The background colour can be anything between a dirtyish brown to a bright golden yellow in sexually active specimens. The fins may show

Localities of *Herotilapia multispinosa*

blueish tones. The pattern of dark markings is also highly variable and its intensity can be linked to certain stages of excitement or behavioural patterns. Besides eight vertical lateral bands a horizontal lateral stripe may also be present, but it is usually broken into

Herotilapia multispinosa

six or seven blotches on the posterior portion of the body. If the fish is in more or less neutral mood, the large lateral blotch in the centre of the flank and the small spot on the caudal peduncle are the only black markings visible. During periods of parental care the entire lower part of the body and the ventral fins turn deep black. Fully grown specimens exhibit a sort of sexual dimorphism that usually allows to distinguish between the sexes. The much larger males mostly grow an indication of a hump on the forehead. Moreover, their anal and dorsal fins are more enlarged and produced.

Cichlid biotope with ample vegetation in Mexico (Rio Pablillo)

Specific traits

that distinguish *Herotilapia multispinosa* from all other American Cichlids and thus make a clear identification of this species possible, are the presence of eleven spines in the anal fin, and several rows of incisor-like, compressed teeth with three tips in both jaws.

Similar species

are found in the genus *Archocentrus*, for example *Archocentrus nigrofasciatus* and *Archocentrus spilurus*. They are, however, unlikely to be mistaken for this Cichlid as they have completely different colour-patterns. The

natural habitats

lie in Central America, and include the Lakes Nicaragua and Managua and the Rio San Juan in Nicaragua as well as rivers on the Atlantic slope of Costa Rica, particularly the drainage systems of the Rio Bebedero, Rio Tempisque, and Rio Frio. Detailed information on the biotopes in Costa Rica was published by BAYLIS (1974). According to this, *H. multispinosa* apparently also lives in vegetated zones of larger water-bodies, but actually seems to prefer isolated pools in their vicinity. These may

partly have temperatures of as much as 34 °C and be as shallow as 25 cm. At sites like these cover was provided by the floating plant, *Azolla mexicana,* and a submerse aquatic species of the genus *Heteranthera*. This biotope is shared with *Archocentrus centrachus, Parapetenia managuensis, Poecila mexicana,* and several Characins. Although *H. multispinosa* is an omnivore, an analysis of stomach contents showed that in nature it largely feeds on algae that it scrapes off solid surfaces. The

care

of this highly adaptable fish is easy. An aquarium with a length of about one metre is spacious enough. It should be decorated with gravel and rocks so that a number of hiding-places are created. Delicate aquatic plants are inadequate as they would be eaten by the fish. Suitable company fishes are such as the smaller Central American *Archocentrus.*

Breeding

can be successful under the most different conditions. Fully grown specimens can produce more than one thousand eggs which like the young fish are cared for in a father-mother family-structure with both parents assuming different responsibilities.

The Genus Hoplarchus
KAUP, 1860

This genus was established by KAUP as early as in 1860. It is monotypical which means it contains only a single species that occurs in northern Brazil, southern Venezuela, and eastern Colombia. The genus belongs to the Cichlasomines and was temporarily dumped into the synonymy of *Cichlasoma*. As a result of the revision and restriction of the genus *Cichlasoma* by KULLANDER (1983), the name *Hoplarchus* has become valid again.

Localities of *Hoplarchus psittacus*

▶ *Hoplarchus psittacus*
(HECKEL, 1840)

Allocated to the genus *Heros* in its original description, this species has since been temporarily regarded as a species of *Acara* (STEINDACHNER) or *Cichlasoma* (REGAN).

Synonyms include *Promotis fasciatus* SCHOMBURGK, 1843, and *Hoplarchus penthacanthus* KAUP, 1860.

Due to its maximum size of more than 30 cm, it is a favourite edible fish in its

Hoplarchus psittacus

home countries and therefore regularly found on fish markets. Influenced by different stages of excitement, its black markings are highly variable. Besides six broad lateral bands there may also be a relatively narrow lateral stripe that usually disintegrates into a horizontal series of spots. The caudal spot lies immediately above the lateral line and is constantly visible. Another dark spot is situated on the posterior lower margin of the eye and on the upper angle of the gill-cover respectively. The iris is conspicuously red. Adult specimens may also display a bright red gular region. This species is isomorphic, i.e. there is no sexual dimorphism that would simplify the determination of the sexes.

Specific traits

by which an identification of this species becomes possible include the shape of the mouth, or rather the head, which resembles the beak of a parrot. Furthermore, there are a cleft in the rear edge of the pre-gill cover not far from its angle, two black spots on the posterior lower margin of the eye and on the upper angle of the gill-cover, and five, more rarely six spines in the anal fin. It can hardly be confused with any other Cichlid as there is nothing like it. The

natural habitats

of this Cichlid are situated in the drainages of the Rio Negro and the upper Orinoco, centring in the border region of the three countries Brazil, Venezuela, and Colombia. This alone already suggests that *H. psittacus* is an inhabitant of typical black waters that are characterized by an extreme deficiency in dissolved minerals and very acidic properties. Despite the aforesaid, the

care

of this fish can also be successful in moderately hard water with a pH above 7. As a

Hoplarchus psittacus considered an edible fish in its home countries.

result of the adult size, the long-term husbandry necessitates very large aquaria of at least one and a half metres in length. To feel at home, the fish should be given opportunities to hide among large roots and/or stone flakes. *H. psittacus* is a tranquil fish which despite its size requires relatively little space for movement. An appropriate diet can be ensured through regular offerings of shrimp and fish meat that, cut into adequate pieces, will be readily accepted after a certain period of adjustment. This fish does obviously not like bright light, but prefers more dim illumination.

Breeding

has apparently been successful in only a few isolated instances and there are no respective reports in the literature. One reason for this situation certainly is that this species is so rarely imported. On the other hand, it grows very large, and with the sexes not being distinguishable with certainty on the basis of external features, it is difficult to assemble potential breeding pairs. All this is complemented by the fact that this fish is highly adapted to the water of its natural environment, very much like in the case of the Discus fish (genus *Symphysodon*) with whom it occurs syntopically. Attempts to breed this species should therefore focus on very soft water with an acidic pH-value and as little nitrogen content as possible.

The Genus Hypselecara
KULLANDER, 1986

Hypselecara is another Cichlasomine genus. It is a very small unit containing only two species of Cichlids from the Orinoco and Amazon regions which were previously assigned to the catch-all genus *Cichlasoma*. Juveniles of either species have fairly pointed heads, but grow a very deep head with a steep forehead as they age.

Many islands in the Rio Negro have white sand beaches.

128

▶ *Hypselecara temporalis*
(GÜNTHER, 1862)

This is one of the few Cichlids kept and bred in German aquaria as early as in the thirties. Originally it was described as *Heros temporalis,* but various subsequent authors erroneously described it again under names such as *Heros goeldii* BOULENGER, 1897, *Acara crassa* STEINDACHNER, 1875, and *Cichlasoma hellabrunni* LADIGES, 1942, which thus are synonyms. Until KULLANDER assigned this Cichlid to his new genus *Hypselecara* in 1986, it was temporarily kept in the genus *Cichlasoma.*

The fish may reach a length of nearly 20 centimetres. The flanks have a greenish yellow to greenish basic colouration that contrasts nicely with the wine-red colour of the fins, the upper and lower portion of the head, the gular, and the chest regions. As an exception, the red colour may even spread over more or less the entire body. The centre of the flanks may be marked with a black spot sandwiched between the lateral lines. The upper half of the caudal peduncle also carries a black spot, and occasionally there is a third one between the eye and the base of the pectoral fin. The intensity of the black markings largely depends on certain moods and stages of excitement.

There is only an indication of a sexual dimorphism with older males developing a steeper forehead and longer dorsal and anal fins.

The most significant

specific traits

that identify *Hypselecara temporalis* include, according to KULLANDER (1983), the number of spines in the anal fin that generally count seven to eight. For aquaristic purposes the wine-red colouration of the body and the fins is far more important.

Localities of *Hypselecara temporalis*

A similar species

which might cause confusion is *Hypselecara coryphaenoides* (HECKEL, 1840) and in the scientific literature both names have repeatedly been used synonymously. *Hypselecara coryphaenoides* has only six spines in the anal fin though and lacks the wine-red colour that is replaced by brown to blackish brown. Moreover, its lateral spot is situated above the upper lateral line (KULLANDER, 1986).

Natural habitats

Hypselecara temporalis inhabits a very large area that stretches over the entire length of the Amazon River. Locality records exist from the central Ucayali in the vicinity of Pucallpa down to the mouthing area (Rio Gurupá). Furthermore, this species has also been found north of the Amazon River in the Oyapock and Amapá rivers. We caught this fish during several excursions through the drainage of the Rio Ucayali in Peru. They appear to favour more or less stagnant water bodies with typical white water which is characterized by low hardness and a neutral to slightly acidic pH.

Hypselecara temporalis

For the care

of a compatible semi-adult pair a tank of at least one metre provides enough space. It should be decorated with stone flakes and roots among which there are a number of hiding-places. It is furthermore recommendable to have some floating plants which dim the light as this species apparently prefers shadowy conditions. If the illumination is too bright, the fish look discoloured and are shy. For the same reason, a dark substrate should be chosen.

Hypselecara temporalis is a calm, relatively peaceful fish which can be kept in company of other South American Cichlids without problems. On reaching a certain size, it requires a rich diet including regular offerings of cut fish and shrimp meat.

Breeding

is not all that easy. Although it was occasionally successful in alkaline, moderately hard water, acidic water poor in minerals undoubtedly provides a more promising basis. It is an openbrooder that forms a parental family-structure. For spawning the partner ignore horizontal surfaces, but choose a vertical site. At a water-temperature of about 25 °C the larvae hatch after approximately three days and are then deposited in pits. They are easily reared on a diet of Brine shrimp and fine flake-food.

Table 27:

Site:	Tributary to the Rio Pacaya south of the village Bretana in the Canal de Puinahua drainage (lower Rio Ucayali, Depto. Loreto, Peru)
Clarity:	relatively turbid
Colour:	brownish
pH:	7.1
Total hardness:	4 °dH
Carbonate hardness:	5 °dH
Conductivity:	127 µS at 26 °C
Depth:	50 to > 200 cm
Current:	hardly recognizable
Water temperature:	26 °C
Air temperature:	26 °C
Date:	3. 8. 1984, 17.00 h

The Genus Mesonauta
GÜNTHER, 1862

Mesonauta species belong to the longest known Cichlids for the aquarium with reports on imports dating from 1908 and first breeding records published in 1911. Originally, the genus contained the three species *Mesonauta insignis* (HECKEL, 1840), *M. festivus* (HECKEL, 1840), and *M. acora* (CASTELNAU, 1855), but GÜNTHER (1862) recognized only *M. insignis* as a valid taxon. This was eventually rectified in part by KULLANDER in 1986.

In 1991, KULLANDER & SILFVERGRIP published another revision of the genus *Mesonauta* as a result of which they not only revalidated all three species, but described two additional ones as well. These authors also highlighted as diagnostic features at species level, the importance of the colour-pattern in general and the black vertical lateral bands in particular. This is very useful for the aquarium keeper as it often allows the identification of live fish on the basis of well visible traits. However, the intensity of these lateral bands is greatly influenced by the prevailing mood of the individual specimen. In general, the banded pattern is more distinct in juveniles than in adult fish.

KULLANDER & SILFVERGRIP stressed that the locality records cited in their paper do not cover the entire distribution, but that significant regions, such as Ecuador, the Guyanas, the central Amazon River, and the lower Rio Negro, were not taken into consideration. Furthermore, they indicated that there are two or three more species of *Mesonauta* which they were unable to describe due to insufficient material available.

The species of *Mesonauta* belong to the generic grouping of the Cichlasomines. They are more closely related to the genera *Pterophyllum*, *Symphysodon*, and *Heros*.

Localities of *Mesonauta insignis*

▶ Mesonauta insignis
(HECKEL, 1840)

This Cichlid was originally described as *Heros insignis* and for subsequent decades considered to be a synonym of *Heros festivus*.

The most significant traits in the colour-pattern of this species consist of an oblique black lateral stripe that runs from the upper lip or the hind edge of the eye, to the posterior portion of the dorsal fin, and a ocellate spot with a golden halo on the upper half of the caudal peduncle. Above the oblique lateral stripe adult specimens show a reticulated pattern of dark tinted scale edges. Fully grown *Mesonauta insignis* display bright yellow to orange yellow areas on the lower head and flanks. The caudal fin and the posterior part of the anal fin are patterned with tiny dark speckles.

Specific traits

This species is characterized by for a banded pattern in which the sixth and seventh band are largely fused or barely separated by a very narrow light line. Moreover, the fifth and sixth band are connected in the ventral region. The intensity of the banded pattern greatly depends on the mood, though, and is usually only visible in juvenile or frightened specimens. Further specific traits include the sort of reticulated pattern above the oblique lateral stripe originating from dark scale edges, and the yellow colouration on the lower body half.

Natural habitats

KULLANDER & SILFVERGRIP (1991) indicated *M. insignis* to be distributed only in the upper Rio Negro and the Orinoco. It is, however, possible that this species also occurs in the lower Rio Negro, in the vicinity of the Anavilhanas Islands for instance.

We were able to catch this species in 1992, in a tributary of the lower Rio Caura that itself flows into the central Orinoco. The locality was a pond-like extension of a larger stream that served as a bathing site. This water-body offered ample hiding-

places in the form of dense swamp vegetation along its margins. In April, at the time of low tide, it was home to a surprisingly large number of Cichlids. Besides *Heros severus* we also recorded *Satanoperca daemon, Mesonauta insignis, Acaronia vultuosa,* and also a species of *Aequidens* and another one of *Apistogramma* from here. The clear water was very soft and distinctly acidic. HANSEN (in litt.), who caught the specimen illustrated here in the Rio Atababo, a tributary to the Orinoco in the border region of Colombia and Venezuela, confirmed our data, i.e. 31-34 °C, total and carbonate hardness < 1 °dH, pH 5.

In comparison with other Cichlids, the ecology of *Mesonautas* is much more focused on the water surface. This is the reason why they can be observed without problems in even very turbid white water. In cases of danger they also would not seek shelter near the bottom like other Cichlids, but rush up to the surface. They often even jump out of the water or hide horizontally in thickets of vegetation.

Mesonauta insignis from the Rio Atabapo

Rio Atabapo, the type locality of *Pterophyllum altum* and *Uaru fernandezyepezi*

One of us had an opportunity to extensively observe *Mesonauta* cf. *insignis* underwater in the clear black water of the Rio Cuieiras in the lower Rio Negro drainage in March 1986. The Mesonautas lived in an about 30 metre wide bay with fine white sand in the company of *Acarichthys heckelii*, *Aequidens pallidus*, *Apistogramma pertensis*, *Biotodoma wavrini*, *Cichla temensis*, *Heros severus*, and *Satanoperca lilith*.

At the time of the observations it was high tide and the vegetation on the banks that consisted of solitary bushes and high grass, was partly submerged for some metres. The Mesonautas were usually seen swimming around near the surface in small groups of three to six specimens. They obviously preferred the immediate vicinity of submerged bushes or branchwork among which they could retreat at a sign of danger. While they hid among the submerged vegetation, they usually showed an intense banded pattern that was almost invisible when the fish resided directly under the water surface near the banks. On such occasions they would assume a position in

which the oblique lateral stripe between the mouth and the hind edge of the dorsal fin appeared horizontal.

Care

Keeping this fish in an aquarium is easy as it is fairly adaptable. Tanks with a length of one metre or more provide enough space

Table 28:

Site:	Banjo of Aripao southeast of the town of Maripa (lower Rio Caura drainage, Venezuela)
Clarity:	> 1 m
Colour:	uncoloured
pH:	5.3
Total hardness:	< 1 °dH
Carbonate hardness:	< 1 °dH
Conductivity:	< 10 µS/cm
Depth:	< 80 cm
Current:	slight
Water temperature:	30.0 °C
Air temperature:	31.5 °C
Date:	10. 4. 1992
Time:	11.00 h

for this calm and peaceful species. When such an aquarium is decorated, it must be ensured that there are a sufficient number of hiding-places between pieces of bog-oak. Some larger plants are crucial as they are preferred spawning sites for this fish. The species of *Echinodorus* have proven suitable for this purpose. A few floating plants provide the privacy required. It should only be kept in the company of Cichlids which are of similar disposition and have comparable ecologies. The species of *Pterophyllum* are highly recommendable as they also occur in the same biotopes. Another choice would be a species of *Symphysodon*, i.e. Discus fish.

Breeding

of specimens of *Mesonauta insignis* imported from the black water rivers of the Rio Negro and Orinoco drainages is not all that easy. The reproductive ecology of Mesonautas is unlike the behavioural patterns of other Cichlids. This includes for example that these fish do not spawn on stones on the bottom, but rather choose leaves of aquatic plants. The larvae are subsequently not deposited in pits, but attached to plants near the water surface. This seems to be opposing the fact that there are no higher aquatic plants in the natural biotopes of these fish. During the rainy season, the home waters of *M. insignis* burst their banks and set the rainforest under water. The water level may then reach such heights that the lower branches of trees are deeply submerged. It is therefore obvious that during such times there is no deficiency in suitable spawning media. It is just that it is no true aquatic vegetation, but submerged terrestrial plants. The fish rear their young in a parental family.

Localities of *Mesonauta mirificus*

▶ *Mesonauta mirificus*
KULLANDER & SILFVERGRIP, 1991

The upward-oblique lateral stripe, so typical of all Mesonautas, and the large ocellate spot on the caudal peduncle with its light halo are well developed in *Mesonauta mirificus*. The area above the lateral stripe is usually fairly dark. This Cichlid is characterized by a pattern of about nine very narrow dark horizontal lines below the lateral stripe that are formed by dark spots on the scales. All unpairy fins and the ventral fins with their whitish extensions are sooty dark grey. The gill-cover and a small area around the lower lip are bright yellow. This species may reach approximately twelve centimetres in length.

Specific traits

Like the other members of this genus, live specimens of this Cichlid can be identified based on the specific arrangement and shape of the dark lateral markings. *Mesonauta mirificus* is characterized by a sixth lateral band whose lower section is particularly broad and bifurcate. This is complemented by a pattern of narrow horizon-

Freshly caught *Mesonauta mirificus* from the Yarina Cocha (Ucayali drainage)

tal lines on the flanks that are made up by each scale having a dark centre. The front lower jaw and the gill-cover of *M. mirificus* are bright yellow.

Natural habitats

The localities known to date, centre in the drainages of the central and lower Rio Ucayali and upper Amazon River up to the town of Letitia, i.e. in Peru and adjacent Colombia. We repeatedly caught this fish in the vicinity of Pucallpa in water bodies that belong to the drainage of the Ucayali. These were characterized by very turbid, moderately hard white water with a pH between 7 and 8.

Unlike other Cichlids *Mesonauta mirificus* is an inhabitant of zones near the water surface where it resides between logs and branches or under the floating carpet of water-hyacinths. Covered areas like these are obviously of great importance for this slow moving and little resistive fish. Its vulnerability is probably also the reason why it

is usually encountered in schools of five to ten specimens outside the reproduction period.

In contrast to other members of this genus, breeding of *Mesonauta mirificus* is relatively easy as the water chemistry is not of particular importance (see chart).

Table 29:

Site:	Isolated pool in the drainage of the Paca Cocha, or Yarina Cocha respectively, near Pucallpa (Peru)
Clarity:	very turbid
Colour:	yellowish brown
pH:	8.0
Total hardness:	13 °dH
Carbonate hardness:	15 °dH
Conductivity:	450 µS at 24 °C
Depth:	> 2 m
Current:	none
Water temperature:	24 °C
Air temperature:	27 °C
Date:	18. 7. 1981
Time:	9.00 h

The Genus Neetroplus
GÜNTHER, 1866

In his 1981 revision, ROGERS allocated only one species to this genus, i.e. *Neetroplus nematopus*, but due to the splitting of the catch-all genus *Cichlasoma*, *Neetroplus panamensis* must also be included in this clade.

▶ *Neetroplus nematopus*
GÜNTHER, 1866

Junior synonyms of this Cichlid include *Neetroplus nicaraguensis* GILL & BRANSFORD, 1877, and *N. fluviatilis* MEEK, 1912. Under the conditions of an aquarium, males can grow to a length in excess of 12 cm while females are considerably smaller.

Localities of *Neetroplus nematopus*

In a neutral mood, the fish are greyish brown to light grey. The flanks show a broad black central band. When the specimens become busy with parental care, a

Neetroplus nematopus in neutral colouration

dramatic colour change ensues. The entire body assumes a very dark, almost blackish colour then and the black band is replaced by a white body-belt that is in great contrast with the colours of the rest of the body. A sexual dimorphism becomes only visible with age. Males not only grow much larger, but also develop more expansive fins.

A

specific trait

that clearly distinguishes *Neetroplus nematopus* from other American Cichlids and thus makes a clear identification possible is the shape of the teeth. In contrast to other Cichlids, the teeth of this species are neither round nor pointed, but flat with a straight apex like incisors. However, for aquaristic needs the typical round, short-headed profile and the colour-pattern should be sufficient as there are not really any similar species.

The

natural habitats

of *N. nematopus* lie in the large Central American rift lakes, i.e. the Lakes Nicaragua, Managua, Kiloa, and Masaya, and in rivers on the Atlantic slope of Nicaragua and Costa Rica (Costa Rica River). This Cichlid is a food specialist which largely feeds on plant material in nature in the form of algae that it scrapes off any solid surfaces. Examinations of stomach contents of specimens from Lake Nicaragua furthermore showed that Copepods (small crustaceans) and larvae of Chironomids (mosquitos) play an important role in the diet. The fry of other Cichlids is also preyed upon.

The

care

of this unusually insensitive and adaptable Cichlid can also be recommended to the

Neetroplus nematopus in breeding colouration

beginner. Chemical properties of the aquarium water are of no concern as higher degrees of hardness and alkaline pH-values meet with the natural requirements of the fish. Feeding them is also not a problem as all common types of food are readily accepted. A compatible pair should be housed in a tank of about one metre in length. When it is decorated, it should not be forgotten that these fish require hiding-places to feel comfortable. Only if there is enough space can it be recommended to keep this species together with other Cichlids. During periods of parental care, *N. nematopus* become very aggressive and usually claim large territories that they defend viciously.

Breeding

is easy. The fish will spawn in a cave if such is available. The brood is cared for in a father-mother family-structure in which the female cares for the eggs and larvae and the male protects the borders of the breeding territory. When the young fish swim free, they can be fed with pulverized flake-food and the larvae of the Brine shrimp.

◗ *Neetroplus panamensis*
MEEK & HILDEBRAND, 1913

Localities of *Neetroplus panamensis*

Due to its peculiar dentition, this Central American Cichlid was originally described as *Neetroplus panamensis*. ROGERS (1981) later transferred it into the collective genus *Cichlasoma* because he believed the shape of the teeth to be not genetically anchored, but rather a result of the fish's ecology and its environment. After the revision of *Cichlasoma* this species now needs to be kept in the genus *Neetroplus* again.

First imports of live specimens into England and Germany were observed in the early eighties, but they were previously kept and bred in the USA. The maximum size of this Cichlid is about 15 centimetres.

The fish is greyish green, rather greenish on the back. Between the base of the pectorals and the caudal peduncle there is a series of irregular black spots on the level of the lower branch of the lateral line. Another series of likewise black spots may be present on the upper parts of the body. Both may be interlinked with one another by lateral bands. The back, nape, and occasionally also the anterior portion of the dorsal fin are covered with tiny silvery to blueish glossy dots that are more obvious in males. There is a red colour-zone immediately behind and above the base of the pectoral fins that may extend over the entire chest

Female *Neetroplus panamensis*

and belly region as far up as onto the dorsal fin. The size and intensity of the red colouration greatly depends on the age of the individual and the present mood.

During periods of parental care the colour-pattern becomes more contrasting. This is especially true for the black markings that then expand. The females in particular at those times show a deep black lower half of the head. Male and female specimens can only be distinguished on the basis of external traits when they become older. Males then develop a steeper forehead, more enlarged fins, and a more intense red colouration.

Male *Neetroplus panamensis*

Specific traits

of this fish are such as the characteristic red and black markings in the colour-pattern, the conspicuously small mouth, and the presence of only nine (more rarely eight or ten) rays in the dorsal and six to eight rays in the anal fin. Chances are little that it is mistaken for any other Central American Cichlid.

With some effort, one could consider as

similar species

some representatives of the genus *Theraps* which also show intense red colours and comparable black markings. These species, however, always have more than ten rays in the dorsal fin.

Natural habitats

Neetroplus panamensis is endemic to Panama with its biotopes lying on the Atlantic as well as on the Pacific slopes. The type locality is the Rio Mandingo in the Canal zone. The distribution of this fish appears to be limited to the drainages of the Rio Chagres and the Rio Tuira here, whereas it is also found in the Rio Sambú and the Rio Chucunaque in eastern Panama

(Darién). SELLICK (in litt.) caught this species in the Rio Frijoles in slightly acidic (pH 6.8) and relatively soft water (conductivity 124 micro-Siemens).

As the fish is fairly adaptable, these values can be neglected as far as

care

or even breeding are concerned. In the long run, the fish should only be kept in an aquarium of more than one metre in length. The decoration of the tank should make provision for a number of cave-like shelters among pieces of bog-oak and larger rocks as such localities obviously are preferred spawning sites. Larger, robust plants will not be damaged by the fish. They can easily be kept together with moderately large Cichlasomines, such as *Archocentrus* or *Thorichthys*.

Breeding

of *Neetroplus panamensis* is easy if an appropriate environment is provided. They preferably spawn on a smooth stone. The eggs, larvae, and young fish are cared for and protected by both the female and the male. Initial feedings should consist of baby Brine shrimp and pulverized flake-food.

The Genus Paraneetroplus
REGAN, 1905

The members of the genus *Paraneetroplus* differ from other Central American Cichlasomines in having teeth of different shapes. All species of *Paraneetroplus* have flattened teeth that lack a second apex.

On occasion of the description of a new species of *Paraneetroplus*, ALLGAYER revised this genus in 1988. Previously it was considered monotypical for decades and temporarily kept in the synonymy of the catch-all genus *Cichlasoma*. ALLGAYER acknowledged only two species to belong to this unit which is opposing the opinion expressed by WERNER & STAWIKOWSKI (1987, 1990) who also allocated *Paraneetroplus gibbiceps* (STEINDACHNER, 1864) and *Paraneetroplus nebulifer* (GÜNTHER, 1860) to this genus.

All representatives of the genus *Paraneetroplus* have a fairly slender build, a convex forehead, and a distinctly low-set mouth with flat teeth that have only one apex. Like the very similar species of the genus *Theraps* they are inhabitants of fast flowing water bodies.

The species of *Paraneetroplus* inhabit waters with a strong current.

◗ *Paraneetroplus bulleri*
REGAN, 1905

Localities of *Paraneetroplus bulleri*

As this Cichlid does not have cylindrical or conical teeth as is the case in the majority of Cichlasomines, it was placed in the separate genus *Paraneetroplus*. Subsequent authors considered this trait to be of minor importance and temporarily kept this species in the genus *Cichlasoma*. Live fish reached Europe for the first time in 1983, thanks to the efforts of STAWIKOWSKI and WERNER. The fish reach a maximum length of nearly 25 centimetres.

The back is olive green to brownish, whitish grey on the belly. The area between the caudal spot and the gill-cover is marked with a series of five black blotches. The posterior two are situated below the soft portion of the dorsal fin and may expand upwards up to the base of the fin. The lower portion of the head show a beautiful red in adult specimens. The pectoral and ventral fins are yellow, the caudal fin and the hind sections of the dorsal and anal fins are reddish. Males and females cannot be told apart with certainty on the basis of external features.

Paraneetroplus bulleri

141

Specific traits

that make a determination of this species possible, include the slender, elongate body shape, a remarkably small mouth, the presence of laterally compressed, flattened teeth, and the colour-pattern in which the five blotches on the flanks are particularly characteristic.

With some justification, *Paraneetroplus nebulifer* (GÜNTHER, 1860) from the Rio Papaloapan, *Paraneetroplus gibbiceps* (STEINDACHNER, 1864) from the drainage of the Rio Teapa in the Mexican state of Tabasco, and *Paraneetroplus omonti* ALLGAYER, 1998, from the Rio Tulija in the state of Chiapas can be considered

similar species.

All these Cichlids have a very elongate, slender build and similar spotted patterns. Despite these superficial similarities, it is quite unlikely to confuse them with *Paraneetroplus bulleri* as there are distinct differences in the colouration. The

natural habitats

of this Cichlid lie in the east of the Mexican state of Oaxaca. It appears to be endemic to a few left tributaries of the upper Rio Coatzacoalcos with the Rio Sarabia being the type locality. We caught this species in the vicinity of the village of Mixtequita, in the Rio Junapan that forms part of the drainage of the Coatzacoalcos, but lies north of the Rio Sarabia. This river diminishes to an about ten metre wide, clear, fast flowing water body at the time of low tide. The depth was nowhere deeper than one metre. The bottom consisted in parts of sand, alternating with sections covered with gravel and large rocks and boulders. These fish were preferably seen where roots of trees, branches, and rocks provided cover. These biotopes were shared with *Parapetenia salvini*, an unidentified species of the genus

Vieja, and *Poecilia sphenops*, a Live-bearing toothed carp. An analysis of the water revealed fairly high degrees of hardness and a clearly alkaline pH. All this suggests that the

care

of *Paraneetroplus bulleri* is not particularly difficult. The fish require large tanks whose length should distinctly exceed one metre. The aquarium should be decorated with washed gravel, large rocks, and roots, which are arranged to form hiding-places and simplify the definition of territories. If plants are to be included, only hardy species can be recommended. The fish does not grow rapidly even if provided with an appropriate varying diet including pieces of meat of freshwater and marine fish, crabs, and shrimp. Until this manuscript was completed, the

breeding

of this species in captivity was an exception. However, there is no reason to doubt that *P. bulleri* is an openbrooder which spawns on stones or roots and that both parents participate in caring for the brood. Initial food offerings for the fry should consist of the larvae of *Artemia salina* and pulverized flake-food.

Table 30:

Site:	Rio Junapan, 9 km from the village of Mixtequita (Coatzacoalcos drainage, Oaxaca, Mexico)
Clarity:	80 cm
Colour:	brownish
pH:	7.75
Total hardness:	15.0 °dH
Carbonate hardness:	11.5 °dH
Conductivity:	1150 µ at 30 °C
Depth:	< 1 m
Current:	strong
Water temperature:	30 °C
Air temperature:	32 °C
Date:	6. 4. 1983, 16.00 h

The Genus Parapetenia
REGAN, 1905

The name *Parapetenia* was originally introduced by REGAN as a subgenus of the genus *Cichlasoma*. The splitting and redefinition of the former catch-all genus *Cichlasoma* by the Swedish ichthyologist KULLANDER in 1983 also resulted in a restriction of this genus to South America so that all previous members from Central America had to be re-allocated to other genera. In the majority of instances there were generic names available that intermittently were considered synonyms of *Cichlasoma*, but had become available again due to said revision. This, however, did not apply to many Cichlids of the section *Parapetenia*. Some modern authors therefore nowadays use the name *Parapetenia* as a valid genus (e. g. ALLGAYER 1989). KULLANDER (1983) had suggested

revalidating the generic name *Nandopsis* GILL, 1862, which applies to a Cichlid in this clade.

The Cichlids originally incorporated in the subgenus *Parapetenia* are unfortunately not of monophyletic origin, but rather represent a heterogeneous assembly that urgently requires a thorough revision. The type species of *Nandopsis* is *Centrarchus tetracanthus* VALENCIENNES, 1831, a species that together with some other Cichlids from Cuba and Haïti forms a clade of their own within the species presently assigned to *Parapetenia*. This group will certainly in future be recognized as a separate genus under the name *Nandopsis*. On the other hand, this is the reason why *Nandopsis* is not applicable to the other species in this group. We therefore follow the suggestion of ALLGAYER to temporarily use *Parapetenia* for these fishes until a revision of this group has been published.

Collecting site of *Mesonauta festivus, Satanoperca papparterra,* and *Heros* sp. in the Rio Guaporé

◗ *Parapetenia bartoni*
(BEAN, 1892)

Localities of *Parapetenia bartoni*

This Cichlid was originally described as a species of the genus *Acara*. Literature records speak of sizes ranging between 20 and 25 centimetres. Our investigations in the natural habitats revealed that females become sexually mature at a length of seven and a half centimetres and males begin to mate at a length of nine centimetres.

During courtship, this fish is certainly one of the most conspicuous Cichlids. Its forehead and the region above the upper lateral line as well as the unpairy fins then assume a silvery white colouration. The remainder of the body becomes deep black except a few individual, glossy, green scales. Save its upper front quarter that is glossy golden, the iris also turns black.

In a neutral mood, the fish are rather inconspicuous. On a greenish grey to beige background colour they show an irregular pattern of small black spots and speckles so that the flanks appear mottled with black.

Most of the time the black pattern is limited to a series of differently sized, very irregularly shaped spots that begin behind the eye and continue up to the caudal spot. Occasionally this series is visible only on the posterior half of the body. Being an isomorphic species, there are no external features that could help to distinguish between the sexes.

Pair of *Parapetenia bartoni* in breeding colouration

144

Specific traits

that make an identification of this species possible, include the unique courtship colouration and the presence of four spines in the anal fin.

Similar species

are mainly such as *Parapetenia labridens* (PELLEGRIN, 1903), a species inhabiting the same biotopes. It is different in having five to six spines in the anal fin and a completely dissimilar courtship colouration. *Parapetenia bartoni* was included in the section *Parapetenia* of *Cichlasoma* by REGAN (1908). The

natural habitat

of this Cichlid lies in the northeast of Mexico, i.e. in the state of San Luis Potosi. Its distribution range is unusually small and confined to the area of the upper Rio Verde that forms part of the drainage of the Rio Panuco. Locality records are somewhat inconsistent with the type locality being Huasteca Potosina, but modern authors believe this species to be endemic to water bodies in the vicinity of the town of Rio Verde today. We managed to observe it there in the Laguna Media Luna and in the Laguna Los Anteojitos in irrigation channels.

Table 31:

Parapetenia bartoni caring for the brood photographed in the natural biotope

The fish inhabit very clear water with extremely hard and distinctly alkaline properties. It was noteworthy that in the Laguna Media Luna this Cichlid had its breeding territories at relatively great depths, i.e. between four and ten metres. In zones near the shore, the fish often resided in the cover provided by water-lilies.

Due to the release of East African Cichlids in its natural habitats, *Parapetenia bartoni* has become a highly threatened species (comp. STAECK & SEEGERS 1984). The

care

of *Parapetenia bartoni* is undemanding as this is a robust and adaptable species. It should not be kept in extremely soft or acidic water, though. An aquarium with a length of about one metre offers enough space to keep a compatible pair. It should be decorated with rocks, roots, and groups of aquatic plants. The fish are voracious feeders accepting all types of commonly used food. Larger specimens should regularly be offered pieces of freshwater or marine fish.

Breeding

is easy. Being openbrooders, the fish spawn on a stone. The eggs, larvae, and young are cared for in a parental family.

Site:	Laguna Media Luna near the town of Rio Verde (San Luis Potosi, Mexico)
Clarity:	about 20 m
Colour:	blueish
pH:	7.9
Total hardness:	53 °dH
Carbonate hardness:	11 °dH
Conductivity:	1680 µS at 30.5 °C
Depth:	30 m
Current:	none
Water temperature:	30.5 °C
Air temperature:	26.0 °C
Date:	24. 3. 1983, 16.00 h

◗ *Parapetenia dovii*
(GÜNTHER, 1864)

Localities of *Parapetenia dovii*

In its original description this Central American Cichlid of the tribe Cichlasomini was included in the genus *Heros*. Due to its size it is of minor importance for the aquarium hobby. It appears that it was first kept in Germany in the early eighties, but in the United States it had its debut much earlier. In nature, the fish grow to a size of between 40 and 50 centimetres. RIEDEL (1965) even indicates a maximum length of 70 centimetres.

The fish are beige to greyish white in colour. Between the elongate caudal spot that usually stretches over the entire height of the caudal peduncle, and the hind edge of the eye, there is a dark lateral stripe with vague borders. This stripe may often be broken to form individual spots. A black streak runs from the lower margin of the eye to the angle of the gill-cover. Another dark spot is situated on the base of the pectoral fins, with a second above the base.

Depending on the mood there may be about eight dark bands on the sides of the body. The fins are beautiful for their green reflections.

A sexual dichromatism and dimorphism become obvious in older specimens. Male fish develop an occipital hump and their fins taper to long points. Their flanks assume a

Parapetenia dovii, female

violet to blueish sheen. The unpairy fins and the body are covered with numerous small black dots arranged in rows on the sides parallelling the scale rows and merging to form a reticulated pattern of black vermicular speckles on the cheeks.

Female specimens lack the spotted pattern, but display the lateral bands more clearly. *Parapetenia dovii* is a polychromatic Cichlid. Besides the normal morph described afore, a golden coloured variety has been found in Costa Rica.

Specific traits

of this Cichlid include the relatively elongate body shape, the long muzzle with the deeply split mouth, and the well developed sexual dichromatism. This is complemented by the presence of twelve rays in the dorsal fin and only six spines in the anal fin.

Similar species

are exclusive to the genus *Parapetenia,* but confusion is only likely to occur with *Parapetenia managuensis* (GÜNTHER, 1869). This species also has a relatively slender body and a large, deeply split mouth. It, however, does not display a sexual dichromatism and has only ten rays in the dorsal, but seven spines in the anal fin.

The

natural habitats

of *Parapetenia dovii* lie in eastern Honduras, Nicaragua, and Costa Rica. In Lake Nicaragua it prefers to reside above rocky ground and is often found in caves. In this lake it was frequently recorded from greater depths. Adult specimens are predators which exclusively feed on other fish. The biotopes of this Cichlid have soft as well as moderately hard water with pH-values ranging from slightly acidic to clearly alkaline levels.

This was a worthwhile day.

For the

care

of this Cichlid it should be taken into consideration that, due to its size, it is actually no fish for the aquarium. Its long-term husbandry will only be possible in extra-large special tanks of at least two metres in length. On the other hand, these fish are fairly inactive. A number of covered places contribute to their well-being. Enabled by a large, stretchable mouth, even semi-adult specimens can take very large food-items and require considerable amounts of food. They should preferably be offered pieces of fish. To date, little is known about a successful

breeding

in captivity. However, it does not appear to be impossible as the fish already mature at a relatively small size. They are openbrooders which prefer a stone to spawn on and rear their offspring in a parental family.

147

▶ *Parapetenia festae*
(BOULENGER, 1899)

Localities of *Parapetenia festae*

This South American Cichlid is a typical representative of the genus *Parapetenia*. Originally, it was described as *Heros festae*. First imports into Europe were observed during the late seventies. Male specimens may grow distinctly longer than 30 centimetres.

Significant traits in the colour-pattern are such as an ocellate spot with a blueish green frame on the upper half of the caudal peduncle, and seven black bands on the flanks. Another two such bands cross the nape and occiput.

Mature fish display a distinctive sexual dimorphism that makes distinguishing between the sexes easy. The males are not only much larger, they also show a greenish yellow background colouration. Their banded pattern may completely be suppressed. Unlike the females, they show numerous blueish green glossy spots on the posterior half of the body that are also

Parapetenia festae, female

apparent on the fins. Female specimens in contrast have an intense orange red to red colouration that stands out from the banded pattern, particularly during periods of parental care. The spinous portion of the dorsal fin and the ventral fins then also assume deep black colours.

For the aquarium hobby the bright orange red body and fin colours are the most significant

specific trait

to properly identify this species. Furthermore, it is characteristic that the ocellate spot lies on the upper half of the caudal peduncle.

Similar species

are found in the genus *Parapetenia* and confusion is likely to occur especially with *Parapetenia urophthalma* (GÜNTHER, 1862). This species may also show bright red colours. Its ocellate spot on the caudal peduncle is, however, much more expansive and covers almost the entire height of the peduncle. Moreover, there are only six bands on the flanks.

Natural habitats

Parapetenia festae appears to be endemic to Ecuador. Localities are known to exist in the drainages of the Rio Guayas, Rio Chanchan, Rio Durango, Rio Clementina, Rio Boba, and Rio Macul. The type locality is the lower Rio Guayas near the city of Guayaquil.

Care

Generally speaking it is not difficult to keep this adaptable and hardy Cichlid as the water chemistry is of minor importance. Problems may be encountered though with regard to its aggression which it exhibits to conspecifics as well as to other Cichlids in too confined surroundings. A compatible

Parapetenia festae, male

pair can be kept in an aquarium of one and a half metres in length if it is decorated with rocks and roots so that shielded sections and a few caves become available. These will enable the inferior specimen to retreat into safety in cases of fights. It is very risky to remove the fry from the caring parents as this often results in severe fights between the parents. Planting the tank is possible, but should be limited to a few single robust plants.

Parapetenia festae has an at least partially predatory ecology. With its very stretchable mouth it feeds on large food-items. Half-grown specimens will therefore need to be fed with nutritious food including regular offerings of pieces of fish and shrimp meat.

Breeding

A compatible pair is a precondition, but if this is available, breeding is not a problem. Accounting for the natural aggression, the pair should be kept in a tank of its own. It is an openbrooder which forms a father-mother family to care for the offspring. While the female attends to the eggs and larvae, the male primarily protects the breeding territory. A larger female is capable of producing several hundreds of eggs. Rearing the numerous juveniles therefore involves frequent partial water changes in order to remove the harmful metabolism products from the aquarium.

149

◗ *Parapetenia grammodes*
(TAYLOR & MILLER, 1980)

Localities of *Parapetenia grammodes*

This representative of the genus *Parapetenia* belongs to the small group of Central American species that were discovered and described only recently. It was first imported into Germany in 1983, by STAWI-KOWSKI & WERNER. The largest known specimen measures nearly 20 centimetres.

The background colour of this fish is brownish to brownish yellow, overlain with a metallic blueish sheen on the lower half of the body. The flanks may be marked with about eight bands and one stripe that extends from the eye to the caudal spot. A second lateral stripe may occasionally appear on the back. A roundish lateral blotch is usually visible on the crossing point of the lower lateral stripe and the fourth band. The cheeks, the nape, and the muzzle are patterned with a number of very fine reddish brown lines. The scales below the upper lateral line have small reddish brown centres. A similar dotted pattern is present on the unpairy fins. These mostly show a metallic blueish green gloss.

External sexual traits that would make distinguishing between males and females possible with certainty, appear to be absent.

Parapetenia grammodes

Specific traits

that differentiate this species from other *Parapetenia* are the relatively slender body and the pattern of fine reddish brown lines on the muzzle, forehead, nape, and the cheeks. The fish is actually unmistakable.

The

natural habitats

of *Parapetenia grammodes* are exclusive to the upper drainage of the Rio Grijalva in the Mexican state of Chiapas and on the western border of Guatemala. Locality records exist, amongst others, from the vicinities of the towns of Villa Flores, Tuxtla Gutierrez, and Comitan (Rio Grande de Chiapa, Rio Sabinal, Rio Salado, Rio Lagartero). The natural biotopes of this Cichlid were found to consist of moderately to extremely hard water with a pH around 7.5 (STAWIKOWSKI 1983). Interesting data on the natural environment were also published by TAYLOR & MILLER (1980) who observed the fish above rocky, sandy, and muddy ground. Except species of *Potamogeton,* the waters were free of aquatic plants, and the Cichlids often resided in the cover provided by branches. Other fishes living in these biotopes were the Cichlid *Vieja hartwegi,* the Characin *Astyanax fasciatus,* Catfishes, Killifishes *(Profundulus* sp.), and Live-bearing tooth-carps *(Poecilia sphenops).*

The

care

of *Parapetenia grammodes* is trouble-free as this Cichlid is robust and has a great ability to adapt. An aquarium with a length of one metre provides enough space. Hiding-places between rocks and roots contribute to the well-being of the fish. Even plants can be used for decorating the tank as this

Cichlide biotope in northern Mexico (Rio Agua-buena, San Luis Potosi)

predatory species does not damage them. As far as feeding is concerned, it should be borne in mind that adult *Parapetenia grammodes* mainly feed on other fish, much like all other *Parapetenia.* Cut freshwater or marine fish, crabs, and shrimp should therefore make up the main portion of the diet.

Breeding

requires hardly any preparation, if only the fish are kept in a suitable aquarium. *Parapetenia grammodes* is an openbrooder which uses stones or roots as spawning media. Both female and male participate in caring for the offspring. The initial diet for the young fish should consist of nauplii of the Brine shrimp and selected small daphnia.

◗ *Parapetenia loisellei*
(BUSSING, 1989)

Localities of *Parapetenia loisellei*

This beautiful Cichlid from Central America has been kept in German aquaria since the late seventies. Despite its large size it was at one stage a very popular aquarium fish in Europe. Until it was formally described in 1989, it was commonly erroneously referred to as *Cichlasoma managuense* or *Cichlasoma friedrichsthalii* (comp. STAWIKOWSKI & WERNER 1985).

Reaching little more than 20 centimetres in length, *Parapetenia loisellei* is the smallest of the group of large predatory Cichlids that are generally called Guapotes in Central America. The most obvious black markings in the colour-pattern of this Cichlid are the light bordered ocellate spot on the upper half of the caudal peduncle, a lateral stripe that is usually broken to form a

series of spots that run from the eye to the caudal spot, and a suborbital band that consists of two separate spots between the posterior edge of the eye and the lower margin

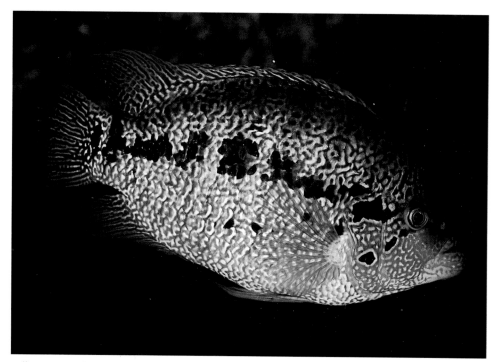

Fully grown male of *Parapetenia loisellei*

Young female of *Parapetenia loisellei*

of the gill-cover. Above the base of the pectoral fin, there is another small black spot.

With adult specimens exhibiting a distinct sexual dichromatism, differentiating between males and females is easy. Females are much smaller and have a bright yellowish background colour. Males in contrast have greenish to greenish golden lower portions of the body and their backs are pale glossy violet. The most obvious difference, however, is the black speckled pattern that adult male specimens show on the lower part of the head, on the flanks, and all unpairy fins.

Similar species

Chances are that this species is mistaken for *Parapetenia friedrichsthalii* (HECKEL, 1840) and *Parapetenia motaguesis* (GÜNTHER, 1866). Both these species have a distinct pattern of black vertical bands on their body sides that is absent in the case of *Parapetenia loisellei*.

Natural habitats

The natural distribution of this Cichlid covers a wide region dominated by the rivers flowing into the Atlantic Ocean, between the east of Honduras and the west of Panama. Individual records also exist from water bodies on the Pacific slopes of Nicaragua, Costa Rica, and Panama, but these may be results of manmade introductions.

Care

Adult specimens can only be kept in very spacious aquaria of at least one and a half metres in length. Rock flakes arranged in such a manner that caves are created make the fish feel comfortable. They require nutritious food including meat of fish.

Breeding

Parapetenia loisellei is a robust openbrooder that has been bred in water with the most different chemical properties.

153

◗ *Parapetenia managuensis*
(GÜNTHER, 1869)

Localities of *Parapetenia managuensis*

Like many other representatives of the genus *Parapetenia* this species was originally allocated to the genus *Heros*. It was first kept in European aquaria in the late seventies. As the fish may reach lengths in excess of 50 centimetres (BARLOW), it is a favourite edible fish in the countries of its natural distribution.

The background colouration of *Parapetenia managuensis* is a silvery whitish grey that sometimes shows a greenish hue. Body and fins are marked with numerous small black spots that may form irregular rows or fuse to streaks. The caudal spot is situated on the upper half of the caudal peduncle. A dark lateral stripe is present, but usually broken into six irregularly shaped blotches. It extends from the edge of the gill-cover to the base of the tail. These blotches may be the start-off points for nebulous bands whose intensity is mood-dependent though. The base of the pectoral fin are usually marked with a black spot. Adult specimens have a brownish red to reddish iris.

As the photograph shows, the lateral stripe, or lateral blotches for that matter, and the lateral bands can be completely absent in some specimens. Moreover, *Parapetenia managuensis* is a polychromatic

Parapetenia managuensis

154

species. Besides the normal morph described above, the natural biotopes also contain specimens of yellow to golden colour. These are, however, very rare. With the sexes being isomorphic, males and females are hard to distinguish on the basis of external characters. Although the spotted pattern is more pronounced in many males, it is no reliable trait to tell the sexes apart.

Specific traits

of aquaristic relevance and that can be used to identify *Parapetenia managuensis* include the deeply split mouth that extends as far back as to below the eye, the pattern of small black spots on the sides of the body, and the absence of a distinct sexual dichromatism.

The genus *Parapetenia* includes a number of relatively

similar species

of which *Parapetenia friedrichsthalii* (HECKEL, 1840) and *Parapetenia motaguesis* (GÜNTHER, 1869) deserve particular mentioning. Both have smaller, less deeply split mouths. *Parapetenia dovii* (GÜNTHER, 1864) has the small black spots on the flanks arranged in regular rows, and they do not fuse to form streaks. This species also has a distinct sexual dichromatism.

The

natural habitats

of *Parapetenia managuensis* lie in Central America. The distribution ranges from eastern Honduras through Nicaragua down to the Atlantic slope of Costa Rica. Locality records exist from the Lakes Nicaragua and Managua, but recently the species has also been recorded from Mexican water bodies where it is likely to have been introduced by man. The fish live in a variety of different types of waters that considerably vary with regard to their chemistry.

Typical clear water rivers have greenish colouration.

As the aforesaid suggests, the

care

of this robust and adaptable Cichlid is easy with regard to the chemical properties of the water. On the other hand, its adult size hardly classifies it as an aquarium fish. Its long-term care will therefore require a special tank of more than one and a half metres in length. Adult specimens are fairly calm and inactive. They prefer to reside in hiding-places between stone flakes. In order to not limit the available swimming space, plants should be used sparsely and limited to a few robust specimens.

Breeding

this large Cichlid in an aquarium is well possible as it already matures on reaching a length of about twelve centimetres. It is an openbrooder which preferably spawns on flat stones. Both parents care for and protect the brood. As they claim large territories during periods of reproduction, they should temporarily be kept in a separate tank. The young fish can be fed with pulverized flake-food and small daphnia.

▶ *Parapetenia* sp. aff. *labridens*
(Tamasopo)

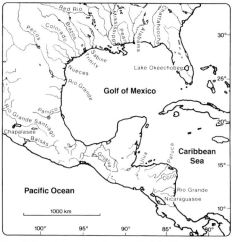

Localities of *Parapetenia* sp. aff. *labridens*

The taxonomic status of the Mexican Cichlid portrayed in the following, at present appears to be doubtful for several reasons. The type locality for *Parapetenia labridens* (PELLEGRIN, 1903) is Huasteca Potosina, Guanajuato, a locality that cannot be properly identified today. TAYLOR & MILLER (1983) included under the name *Parapetenia labridens* the Cichlid from the Rio Aquabuena near Tamasopo portrayed here as well as a species occurring in the Laguna Media Luna near Rio Verde. Our own research in Mexico revealed that these two fishes are distinctly different. This could mean two things, i.e. either *Parapetenia labridens* is a highly variable species, or the name, as it is presently understood, in fact includes two species. We tend to support the second opinion.

The Cichlid depicted in the following was first imported into Germany by ourselves in 1983. The specimens we observed in nature had a maximum length of about twelve centimetres.

In a neutral mood, they show a very irregular black lateral stripe on a whitish grey background colouration. This stripe runs from the upper lip below the lower half of the eye up to a dark spot on the caudal peduncle. It is usually broken into a series of

Parapetenia sp. aff. *labridens*

individual blotches that occasionally form approximately seven vertical bands on the posterior half of the body. During periods of parental care the head, throat, and chest below the lateral stripe assume a deep black colouration. This colour-zone is quite sharply defined and ends at the level of the base of the ventral fin. Like a chessboard, the posterior half of the body above the lateral stripe is also coloured black from about the seventh spine of the dorsal fin backwards. An attractive contrast is created by the upper front and lower hind parts of the body showing a pale shade of beige.

Distinguishing between the sexes on the basis of external traits is difficult, but males generally grow larger. The largest males we observed in the natural biotopes measured about 12 centimetres. A primary

specific trait

that allows an identification is the chequered black and white colour-pattern displayed during breeding. This is complemented by the presence of 16 spines in the dorsal fin, with any lower number being a great exception. A

similar species

with comparable morphometric traits that is presently also considered to represent

Table 32:

Site:	Rio Aguabuena near Tamasopo (San Luis Potosi, Mexico)
Clarity:	5 m
Colour:	green
pH:	9.15
Total hardness:	56 °dH
Carbonate hardness:	9 °dH
Conductivity:	1550 µS at 28.5 °C
Depth:	< 3 m
Current:	partly stron
Water temperature:	28 °C
Air temperature:	30 °C
Date:	23. 3. 1983, 17.00 h

Parapetenia labridens (PELLEGRIN, 1903) originates from the Laguna Media Luna near Rio Verde. This Cichlid, however, differs in having a greenish yellow colouration which turns bright lemon yellow during breeding. These specimens usually have 15 to 16, sometimes even 17 spines in the dorsal. REGAN (1908) included *Parapetenia labridens* in the section *Theraps* of *Cichlasoma*, but today it is considered a member of *Parapetenia*.

The

natural habitat

of this Cichlid is the Rio Aguabuena in the drainage system of the Rio Panuco. We caught specimens near the cataracts of Tamasopo, in the Mexican state of San Luis Potosi. The locality is a typical clear water river with a fast flow in certain sections and mostly a bed of stones and rocks. The water was extremely hard and alkaline. For breeding, the fish chose calmer sections of the water body where the bottom was sandy or muddy. They shared their habitat with *Parapetenia steindachneri*, JORDAN & SNYDER, 1899, another Cichlid of the genus *Herichthys*, a Swordtail *(Xiphophorus montezumae)*, a Live-bearing toothcarp *(Poecilia mexicana)*, and the Characin *Astyanax fasciatus*.

The appropriate

care

of this Cichlid requires an aquarium of at least one metre in length. It should be decorated with fine washed gravel, larger rocks that provide caves, and some bog-oak. Plants can be used as the fish will not damage them. Suitable co-inhabitants for such tank can be found among other small Mexican Cichlasomines. Feeding this Cichlid is not a problem since it readily accepts all common types of food inclusive of flake-food.

No reports on the

breeding

of this Cichlid were available by the time this manuscript was finalized. It should not be a problem though, taking into consideration the water chemistry and environmental conditions of the natural biotopes. Our underwater observations revealed that this Cichlid spawns in caves that it apparently digs itself under flat rocks. The brood is cared for and protected by both parents.

Parapetenia sp. aff. *labridens* and *Herichthys tamasopoensis* in their natural habitat

◗ *Parapetenia trimaculata*
(GÜNTHER, 1869)

Localities of *Parapetenia trimaculata*

This Cichlid was originally placed in the genus *Heros* by GÜNTHER. Junior synonyms are *Cichlasoma mojarra* MEEK, 1904, *C. centrale* MEEK, 1906, *C. gordonsmithi* FOWLER, and *C. cajali* ALVAREZ & GUITIERREZ. The recorded maximum length is 35 centimetres.

Fully coloured specimens exhibit an intensely rosy to reddish colour-zone that stretches over the gular and pectoral regions and may sometimes extend upwards to below the lateral line. The iris is blood-red while the body is yellowish green, and the head is yellowish. The upper portion of the base of the caudal fin is marked with an ocellate spot with a green glossy halo. Another spot with such surrounding is very conspicuous on the shoulder above the base of the upper lateral line. Two or three such markings of much smaller size may be present immediately behind the rear edge of the eye. Depending on the mood, the flanks can sometimes show seven dark bands, the posterior five of which including a deep black mid-lateral spot each. These spots form a horizontal row on the posterior part of the body. With a distinct sexual dimorphism being absent, the sexes cannot be distinguished with certainty on the basis of external features.

Parapetenia trimaculata

159

Cichlids preferably reside in calmer sections of rivers that have strong currents.

Specific traits

of this species are the red colour-zone on the throat and chest, the black shoulder spot, and the small spots behind the eye. These characteristics are sufficient to avoid confusion with any other, superficially similar species. REGAN (1905) included *Parapetenia trimaculata* in the genus *Cichlasoma*.

The

natural habitats

of this Cichlid are exclusive to the Pacific slope of Central America. They lie in the Mexican states of Oaxaca and Chiapas, and in the countries Guatemala and El Salvador. The northern border of the range is said to be the Laguna de Coyuca near Acapulco, the southernmost locality is the Rio Lempa. We managed to record the fish from the Isthmus of Tehuantepec east of the town of Juchitan in 1984. There, it lived in a small stream of about half a metre in depth with a considerable current. The fish preferably resided in the cover provided by overhanging terrestrial embankment vegetation or among reed. The water was relatively turbid, moderately hard, and of an alkaline quality. The biotope was shared with other Cichlid species which we did not identify more precisely.

As far as the

care

of this fairly adaptable species is concerned, the water chemistry need not be particularly worried about. Problems may be encountered with regard to the space requirements though. This Cichlid may become quite aggressive towards conspecifics, and also towards other Cichlids, and should therefore be only kept long-term in tanks of about one and a half metres in length. In order to simplify the definition of territories for the fish, the ground space should be structured with roots and rocks so that there are blinds and hiding-places.

Co-inhabitants of such tank must be equally resistive and strong. Larger specimens require nutritious food and should regularly be offered cut fish meat.

Under the provision that there is a compatible pair available,

breeding

is easy. *Parapetenia trimaculata* is a typical openbrooder which usually spawns on a flat stone. The hatching larvae are then deposited in a pit. Both female and male care for the offspring in more or less equal shares. Rearing the young fish by initially feeding them with pulverized flake-food and baby Brine shrimp is easy.

Table 33:

Site:	Small river between Juchitan and la Ventosa (Oaxaca, Mexico)
Clarity:	relatively turbid
Colour:	brownish
pH:	7.75
Total hardness:	15.0 °dH
Carbonate hardness:	11.5 °dH
Conductivity:	1150 µS at 31 °C
Depth:	< 150 cm
Current:	swift
Water temperature:	31 °C
Air temperature:	34 °C
Date:	6. 4. 1983, 11.00 h

◆ *Parapetenia urophthalma*
(GÜNTHER, 1862)

Localities of *Parapetenia urophthalma*

This Central American Cichlasomine was allocated to the genus *Heros* when first described. A synonym is *Heros troscheli* STEINDACHNER, 1867. This species had already been kept in German aquaria before the First World War, but soon disappeared and was forgotten about until it was imported again in 1980. The maximum body length of this fish may be more than 20 centimetres.

This Cichlid has a yellowish brown to greyish brown basic colouration and shows six dark bands of varying intensity between the bases of the pectoral and the caudal fins. The caudal peduncle is marked with a large ocellate spot in a blueish green surrounding that covers almost the entire height. Particu-larly in juveniles, the upper head region and the throat may show reddish tones. The caudal fin and the rayed portions of the

Parapetenia urophthalma

dorsal and anal fins usually also show pale red tinges. With the sexes being isomorphic, they are hard to distinguish based on external characteristics alone. The only hints are provided by the fact that older males have larger fins.

As *Parapetenia urophthalma* includes several geographical subspecies, deviations from the above colour description are well possible.

Specific traits

of this Cichlid include the large ocellate spot on the caudal peduncle that covers almost the entire height and the presence of six dark lateral bands. Furthermore, it is characterized by 16 (15 or 17 are rare exceptions) spines in the dorsal and six (very rarely five or seven) spines in the anal fin.

Natural habitats

Parapetenia urophthalma inhabits a vast area with localities being known from a fairly wide variety of water bodies on the Atlantic slope of Central America. The distribution of this fish ranges from the drainage of the Rio Coatzacoalcos in the Mexican state of Veracruz, through the Yucatán Peninsula, Guatemala, Belize, and Honduras, up to Nicaragua. The type locality is the Laguna de Petén in Guatemala. HUBBS (1936) distinguished between altogether nine subspecies, seven of which occur in Mexico, i.e. *Cichlasoma urophthalmus alborum* from the Rio Usumacinta (state of Tabasco), *C. u. aguadae* from the vicinity of Tuxpena (Campeche), *C. u. cienagae* from the surroundings of Progreso (Yucatán), *C. u. amarum* from the Isla Mujeres, *C. u. conchitae* and *C. u. zebra* from areas north of the town of Meridan (Yucatán), *C. u. maxorus* from the vicinity of Chichen Itzá (Yucatán), *C. u. trispilus* from the Rio San Pedro (Guatemala), and *C. u. stenozonus* from a locality not precisely known. Occasionally, this

Cichlid was not only recorded from freshwater, but also from brackish environments.

Care

The water chemistry is of minor importance for successful keeping in the aquarium. The tank for these fish should be longer than a metre in length. It is decorated with stone flakes and larger pieces of bog-oak in such a way that hiding-places are created, and the ground space is well structured, so to enable the specimens to define individual territories. Although this Cichlid does not feed on plants, vegetation should be limited to a few large single plants.

The stretchable, deeply split mouth of this Cichlid is an indication of its predatory feeding habits, and adult specimens also feed on fish. This should in so far be accounted for that pieces of fish and shrimp are made regular elements of their diet.

If you have a compatible pair,

breeding

is not a problem. *Parapetenia urophthalma* is a typical openbrooder which uses smooth stones and roots as media to spawn on. The eggs, larvae, and young fish are cared for and protected by both the female and the male. The fry can easily be reared on an initial diet of baby Brine shrimp and pulverized flake-food, followed by selected small daphnia after a few days.

The Genus Petenia
GÜNTHER, 1862

was suggested by GÜNTHER in 1862. It accommodates only one species that is closely related to the Cichlids assigned to the genus *Parapetenia*.

The generic name refers to one of the collecting sites of the fish, the Lake Petén in Guatemala.

▶ *Petenia splendida*
GÜNTHER, 1862

Localities of *Petenia splendida*

Reaching a maximum body length of about half a metre and a weight of about two and a half kilograms, this Cichlid ranks among the largest representatives of the family Cichlidae. It is therefore no surprise that it is one of the most popular edible fish in Central America regularly offered on fish markets.

The basic colouration of this fish is greyish white to pale yellowish. Head, body, and all unpairy fins are covered with numerous small black dots and dashes. One of the most prominent markings is an ocellate spot with a golden surrounding that lies on the upper half of the caudal peduncle. Another black spot is found on the upper angle of the gill-cover. These both are linked by a horizontal series of five black blotches. The area of the throat, but also the forehead and nape, may show a beautiful rosy red colouration. This colour may be repeated on the caudal fin and as a narrow marginal stripe on the dorsal fin. *Petenia splendida* belongs to the small group of

Petenia splendida

American Cichlids which exhibit a poly-chromatism. Besides the normal colour-pattern described above, an orange yellow colour morph is occasionally found. It has a whitish lower half of the body, and the back, forehead, and head are intensely orange. The head is covered with small red spots.

The fish is isomorphic which means there are no external traits that could be used to distinguish between the sexes. The

natural habitats

of this Cichlid lie in the Mexican states of Veracruz, Tabasco, Chiapas, and Campeche, and in Guatemala (Laguna de Petén)

While diving with mask and snorkel in the Rio Nututun (Chiapas, Mexico), we could observe *P. splendida.*

and Belize. We recorded it in 1984 amongst others from a tributary to the Rio Papaloapan that probably represents the western border of its distribution. This Cichlid appears to be fairly adaptable as it occurs in the very turbid lower sections of larger rivers, as well as in clear mountains streams with a strong current. While diving in a pond-like backwash of the Rio Nututun near Palenque, we observed an adult pair that was guarding a school of several hundred young fish. This small river was comparable with Central European trout waters and was also inhabited by at least another five different Cichlids including representatives of the genera *Thorichthys, Theraps,* and *Parapetenia.* Furthermore, we recorded the Characin *Astyanax fasciatus,* the Livebearing toothcarp *Poecilia sphenops,* a Silverside species, a Catfish, and a Sleeper. All localities where we found *P. splendida* had moderately hard, alkaline water.

Due to its size, this Cichlid is actually not a fish for the aquarium, and its

Petenia splendida is a popular edible fish.

care

should rather be left to zoological gardens and museums with their huge display tanks. There, it makes an impressive and interes-

ting display and is easily kept thanks to its adaptability and robustness.

Since adult specimens are voracious predators which preferably feed on fish, they should regularly be offered pieces of freshwater or marine fish. The aquarium should also contain some covered spots among roots as this obviously contributes to the fish feeling at ease.

Breeding

this Cichlid has so far been a very rare exception that certainly has to do with its size. Observations made in nature have revealed that *Petenia splendida* is an open-brooder which rears its offspring in a parental family. Both the female and the male care for and protect the school of young fish for a relatively long period.

Table 34:

Site:	Rio Nututun near Palenque (Chiapas, Mexico)
Clarity:	ca. 3 m
Colour:	greenish
pH:	8.85
Total hardness:	12.5 °dH
Carbonate hardness:	12.0 °dH
Conductivity:	1200 µS at 27 °C
Depth:	150 – 350 cm
Current:	strong
Water temperature:	27 °C
Air temperature:	30 °C
Date:	2. 4. 1983
Time:	14.00 h

The Genus Pterophyllum

was introduced by HECKEL in 1840. A junior synonym is *Plataxoides* CASTELNAU, 1855. It is a small Cichlid genus that presently holds only three species. Typical features of these Cichlids are the laterally greatly compressed, almost disc-like body, the enlarged, filamentous ventral fins, and the minute scales. Close relationships exist on the one hand with the genus *Symphysodon* HECKEL, 1840, and on the other hand, with the genus *Mesonauta* GÜNTHER, 1862.

Localities of *Pterophyllum altum*

▶ Pterophyllum altum
PELLEGRIN, 1903

In contrast to the Common Angelfish, *Pterophyllum scalare* (LICHTENSTEIN, 1823), which was kept in German aquaria as early as in the first decade of the twentieth cen- tury, the Giant Angelfish was first imported not before the early fifties. It can reach a standard length, i.e. the body length with

Pterophyllum altum

out the caudal fin, of 10 cm. The profile line of *Pterophyllum altum* between forehead and nape shows a characteristically sharp indentation at the level of the eye. Furthermore, it is characterized for the second lateral band, that connects the spinous portion of the dorsal fin with the anal region, considerably tapering downwards. In contrast to the other representatives of this genus, in which the interspaces between the four distinct vertical bands are broken by indications of short streaks, the Giant

Angelfish has complete bands. These are less intense in colour, but proceed up to the lower edge of the body. The banded pattern of the fins and the spots on the back are often brownish red or even reddish in this species. It is isomorphic, i.e. the sexes cannot be determined with certainty on the basis of external characters.

Specific traits

Contrary to common belief, the indentation above the eyes is no reliable trait to identify

Pterophyllum altum is a typical inhabitant of black water biotopes that are transparent, but have the colour of tea.

this Cichlid as it is a phenomenon also present in a morph of *Pterophyllum scalare* from the central Ucayali in Peru. An undoubted determination is, however, achieved by counting the row of scales below the lateral line and the rays of the dorsal and anal fins. *Pterophyllum altum* has 46-48 scales in one row on the flanks, 28-29 rays in the dorsal, and 28-32 rays in the anal fin.

Similar species

As the three species of *Pterophyllum* have more or less identical body-shapes and colour-patterns, they cannot really be distinguished on the basis of these traits alone. A reliable identification requires counting the scales in one horizontal row and establishing the number of rays in the anal and dorsal fins. *Pterophyllum scalare* (LICHTENSTEIN, 1823) has 30-40 scales in one horizontal row, 23-29 rays in the dorsal, and 24-28 rays in the anal fin. *P. leopoldi* (GOSSE, 1963) has 26-30 scales, 29-32 rays in the dorsal, and 19-22 rays in the anal fin and lacks the indentation in the forehead profile.

The

natural habitat

of this Cichlid is relatively limited. All localities known to date lie in the drainage of the upper Rio Negro and Orinoco. The type specimen was collected in the Rio Atabapo. *P. altum* lives in waters characterized by an unusually low content of dissolved minerals and an extremely low pH. It appears that the fish highly depends on such environment at least during the breeding season, very much like in the case of the Discus fishes of the genus *Symphysodon*. They are preferably seen standing among the branches of submerged trees in calm river sections with a minor current.

When choosing an aquarium for these fish, the particular shape of the body must be taken into consideration. A successful

care

thus calls for a relatively high aquarium of at least 50 cm in height. It should be decorated with bog-oak and long-stemmed plants so that the fish find secure zones between branches and plants where they can reside. Angelfishes are very peaceful aquarium fishes which cannot be kept together with very agile or even aggressive species. They are, however, most suitable as company for Discus fishes for these have similar behavioural patterns.

Although these Cichlids can also be kept in harder water with alkaline properties, their

breeding

seems to necessitate soft and acidic water. So far, it has been a great exception to prompt this species to reproduce in captivity. Experiences made during the first successes (e.g. LINKE 1994) suggest that this species, like the other representatives of the genus, spawns on leaves and roots high above the bottom and rear their offspring in a parental family-structure.

The Genus Satanoperca
GÜNTHER, 1862

Satanoperca belongs to those Cichlid genera that were already described in the previous century, but then were kept as synonyms for a long time. KULLANDER revived it in 1986 when he extracted a group of so-called Eartheaters that were at that stage included in the then polyphyletic genus *Geophagus*.

This genus at present holds half a dozen of species which are complemented by a number of undescribed forms (KULLANDER & NIJSSEN 1989, KULLANDER 1994). Their distribution extends over the drainage of the Amazon River, the Orinoco, and the Guyanas. A single species, *Satanoperca pappaterra,* also occurs in the Rio Paraguay. It is likely that more splitting of the genus is to follow as it is still not a monophyletic assembly (KULLANDER 1994).

A colour trait differentiating all species of *Satanoperca* from other Eartheaters of the tribe Geophagini, is a black spot usually bordered light that these Cichlids carry on the upper half of the base of the caudal fin. The genus contains substrate-brooders as well as larvophile mouthbrooders.

▶ *Satanoperca acuticeps*
(HECKEL, 1840)

This Cichlid was originally allocated to the genus *Geophagus*. Because of its appearance and behaviour, it may be considered a typical Eartheater. The background colouration is greyish green to greenish golden with most scales on the flanks having a green glossy centre. The iris is mainly red. A black band, bordered greenish golden, runs from the lower edge of the eye to the angle of the mouth. The sides of the body are marked with three black blotches that represent an important identification aid. The first blotch approximately lies below the third to the sixth spine of the dorsal fin, and the remaining two at the level of the last spines and posteriormost rays respectively.

Table 35:

Site: Lago Janauacá southwest of the city of Manaus (central Amazon drainage)	
Clarity:	turbid white water
Colour:	loam-yellow
pH:	7.2
Total hardness:	< 1°dH
Carbonate hardness:	< 1°dH
Conductivity:	52 μS/cm
Depth:	> 2 m
Current:	none
Water temperature:	27°C
Air temperature:	30°C
Date:	20. 3. 1986, 14.00 h

Localities of *Satanoperca acuticeps*

A fourth spot is situated on the upper base of the caudal fin. The anal fin, the lower half of the caudal fin, and the conspicuously enlarged ventral fins are pale yellowish or orange in colour.

Fully grown specimens of *Satanoperca acuticeps* are characterized by for their high dorsal fins whose second to fifth ray are extremely enlarged and end in filaments. The membranes between the anterior spines are also unusually large. Identification of the sexes is difficult as there is no distinct sexual dimorphism. The maximum length of this species lies at about 20 centimetres.

Specific traits

The most significant determination criterium, unique among the Eartheaters, is the presence of three blotches on the flanks. Another specific feature is the high dorsal fin with its membranes that are particularly produced in the anterior portion of the fin.

Similar species

Satanoperca daemon has two, *Satanoperca lilith* has even only one lateral blotch. Other resembling members of this genus lack lateral blotches altogether.

Natural habitats

The distribution of *Satanoperca acuticeps* ranges from the central and lower Amazon River, through the lower Rio Branco, to the Rio Negro in northern Brazil. We repeatedly recorded this species from the wider vicinity of Manaus in the drainages of the Rio Solimoes and Rio Negro. These typical white and black water rivers usually were of very soft and extremely acidic quality, but occasionally also had slightly alkaline properties. During the periods of high water, the fish resided in the cover of submerged embankment vegetation or among twigs and branches lying in the water. Other Cichlids found in these biotopes were *Acarichthys heckelii* and representatives of the genera *Apistogramma, Cichla, Heros, Aequidens,* and *Mesonauta.*

Care

The long-term husbandry of this Cichlid will only be successful in a spacious aquarium of at least one and a half metres in length. The bottom needs to be covered with fine sand as this species, like all Eartheaters, has an instinctive urge to chew through the substrate in search of food. A few large pieces of bog-oak are important as they simplify the territorial division of the available space and provide sanctuaries for the fish. Some larger plants may also be used for the decoration as *Satanoperca acuticeps* will not damage them. Their roots should, however, be covered with stones so that the fish cannot dig them out when they search for food.

This Cichlid must be termed extremely peaceful and shows very little resistance towards other fishes. Suitable co-inhabitants of a community tank must therefore be Cichlids with similar behavioural patterns. Species of the genera *Geophagus, Biotodoma, Pterophyllum,* and *Symphysodon* are particularly adequate for this purpose.

Breeding

Although *Satanoperca acuticeps* is regularly imported in a few, usually single specimens, yet there is nothing known about a successful reproduction in captivity. This is the reason why there is no information available on the breeding behaviour of this species. According to the observations made in the natural biotopes, the most promising preconditions would certainly include very soft water with an acidic pH.

Satanoperca acuticeps

⬧ *Satanoperca daemon*

(HECKEL, 1840)

Localities of *Satanoperca daemon*

is a rarely kept large Cichlid. Imports of this species are infrequent, and captive breeding has only been successful in a few isolated cases. This species may attain 30 centimetres in length.

Indicative traits of the colour-pattern are such as the two black lateral spots, one lying approximately in the centre of the body and the other on the posterior half of the body, i.e. in the area between the soft portions of the dorsal and anal fins. Moreover, there is a caudal spot with a golden halo on the upper half of the base of the caudal fin. The flanks are patterned with about eight horizontal rows of greenish golden glossy spots. The lower half of the caudal fin, and also parts of the anal and ventral fins are brownish red to orange red. This species is characterized by its extremely enlarged ventral fins and the filamentous appendices of the first three or four rays in the dorsal fin. *Satanoperca daemon* is isomorphic, which means there is no distinct sexual dimorphism that would allow an identification of the sexes.

The

specific traits

include the elongate body shape, the filamentous prolongation of the ventral fins and the anterior rays of the dorsal fin, and the three black spots on the sides of the body. Another diagnostic feature is the relatively high number of 18-22 gill-rakers.

Satanoperca daemon

171

Confusion with other Cichlids is quite unlikely to occur as there are hardly any

similar species.

Acarichthys heckelii also has the first rays of the dorsal fin enlarged, but otherwise is much deeper in its build and lacks a caudal spot.

Satanoperca acuticeps (HECKEL, 1840) has three lateral spots in addition to the caudal spot. *Satanoperca lilith* KULLANDER & FERREIRA, 1988, is the only Cichlid which really resembles *Satanoperca daemon* in all aspects, but this species lacks the lateral spot on the posterior half of the body. The

natural habitats

of *Satanoperca daemon* are black and white water rivers characterized by extremely low contents of dissolved minerals and usually a very acidic quality. The distribution extends over the drainage systems of the Orinoco and upper Rio Negro. For the

care

of this Cichlid, the aforesaid imposes that it should preferably be kept in soft water with a pH clearly lower than 7. Due to its final size, its long-term husbandry will necessi-

tate an aquarium with a length distinctly over one metre. Some hiding-places in the form of pieces of bog-oak, rock constructions, or groups of plants, in conjunction with dim illumination create a comfortable environment for this calm, sometimes even a little shy fish. Other Geophagines appear to be the best suitable co-inhabitants of such a tank. It is imperative that the aquarium is not furnished with gravel, but sand, as it is an instinctive behaviour of *Satanoperca daemon* to chew through the substrate in search of food. Unfortunately, this fish appears to be very susceptible for hexamitosis.

Breeding

of this Cichlid has only been successful in a very few isolated cases and the information available on its reproductive behaviour is therefore incomplete. The most favourable conditions for breeding this species will certainly include nitrate-free, acidic water poor in minerals. It is possible that this fish is as highly adapted to its natural environment as the Discus fish which inhabits the same water bodies. If the water quality differs from that, the fish may be difficult or even impossible to breed. It is a substrate breeder which rears its offspring in a parental family (ECKINGER 1987).

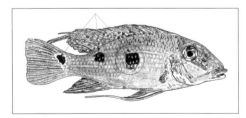

Satanoperca daemon

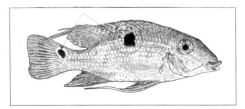

Satanoperca lilith

Table 36:

Site:	Banjo of Aripao, a few kilometres southeast of the town of Maripa (Rio Caura drainage, Venezuela)
Clarity:	> 1 m
Colour:	uncoloured
pH:	5.3
Total hardness:	< 1 °dH
Carbonate hardness:	< 1 °dH
Conductivity:	< 10 µS/cm
Depth:	< 80 cm
Current:	slight
Water temperature:	30.0 °C
Air temperature:	31.5 °C
Date:	10. 4. 1992, 11.00 h

The Genus Symphysodon
HECKEL, 1840

was suggested by HECKEL in 1840. The systematics of this fairly small genus are presently subjected to controversial discussion. Generally, there are two major forms which some authors consider full species — an opinion followed here. Others think of them as two subspecies of the same species. An extraordinary controversy exists with regard to the validity of another four subspecies suggested by SCHULTZ (1960) and BURGESS (1981). These appear to be interlinked with one another through intermediate forms which make clear definitions difficult.

The genus is caracterized by the laterally greatly compressed, deep, almost disc-shaped body. Furthermore, the rays in the dorsal and anal fins are unusually numerous, and there are high counts of the small scales. The dorsal fin contains 8-10 spines and 28-33 soft rays, the anal has 6-9 spines and 27-31 soft rays. A horizontal scale row consists of 48-62 scales (comp. KULLANDER 1986). The genus is on the one hand closely related to *Pterophyllum* HECKEL, 1840, and on the other hand to *Heros* HECKEL, 1840.

While *Symphysodon discus* has only a limited distribution in the rivers Trombetas, Abacaxis, and Rio Negro, *Symphysodon aequifasciatus* inhabits an expansive area in the Amazon region. KULLANDER (1986) indicates the vicinity of Belém to be the eastern border of the range and water bodies in northeastern Peru as the western limit. We caught this species in the Rio Nanay southwest of the city of Iquitos in 1990.

Discus fish belong to the Cichlids most intensely used commercially with many professional breeders worldwide doing nothing else but breeding Discus fish. Cross-breeding and selective pairing of specimens and colour morphs have resulted in a multitude of varieties whose colourations often hardly resemble the free-ranging fish.

Localities of *Symphysodon discus*

▶ Symphysodon discus
HECKEL, 1840

In the aquarium trade, this Cichlid is referred to as the "Heckel's Discus" or "True Discus". First live specimens arrived in Europe in the 1920s. The species may reach a total length of almost 20 cm. The fish is characterized by the fifth of the altogether nine lateral bands between the eye and the base of the caudal fin, being particularly wide and intensely coloured. The flanks are patterned with reddish to brownish red and blue, glossy, undulating stripes. Distinguishing between the sexes on the basis of external traits is impossible as the fish are isomorphic.

Specific traits

that make this species differ from other representatives of the genus, include the prominent fifth lateral band and the pattern of reddish and blue undulate lines that mark the flanks horizontally. Another feature of *Symphysodon discus* is the presence of only 48-56 scales in a horizontal row.

Turquoise Discus fish, a select breed of *Symphysodon aequifasciata*

Similar species

With some justification, one must mention *Heros severus* here, as juveniles of both species have a number of traits in common. In contrast to adult Discus fish, this species, however, has pointed dorsal and anal fins and a less circular body shape.

Symphysodon aequifasciatus PELLEGRIN, 1903, the second representative of the genus *Symphysodon,* differs from Heckel's Discus in so far that the nine dark lateral bands between the eye and the base of the caudal fin are all of equal colour intensity. Furthermore, these fish have 54-62 scales in one horizontal row.

The natural habitats

of Heckel's Discus lie in the area drained by the Rio Trombetas and lower Rio Negro. More recently (BURGESS 1981), this Cichlid was also discovered in the Rio Abacaxis, a tributary to the Rio Madeira. The fish live primarily in typical black water with a pH ranging generally between 5.0 and 6.5. This water is so poor in dissolved minerals that its total as well as its carbonate hardness never reach 1 °dH. The conductivity is always below 20 micro-Siemens.

Preferred sites for this Cichlid are slow-flowing sections of rivers, such as backwaters and tranquil bays with plenty of sub-

174

merged wood. The fish obviously love to reside among branches or roots of drifting islets or floating patches formed by various types of grass. Outside the breeding season, the Discus fish does not live in pairs, but in schools containing as many as 50 individuals.

If some basic rules are followed, the successful

care

of Discus fish is less complicated than generally believed. Soft and acidic water is not necessarily a precondition for keeping this fish healthy for many years. Much more important than the hardness of water and the concentration of hydrogen-ions is the quality of the water with regard to nitrite and nitrate. In order to prevent these nitrogen-compounds from reaching harmful levels, regular partial water exchanges become an absolute must as this is the only way to reduce the ever accumulating concentration metabolism products.

Discus fish can be kept in well planted aquaria as they do not even damage the most delicate vegetation. Covered sites among roots or long-stemmed plants into which the fish may retreat are imperative if they are to feel at ease. They are very peace-

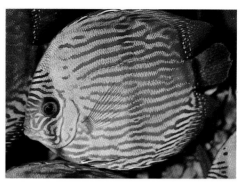

Symphysodon discus

ful and must never be kept together with more active or even aggressive fishes. This is the reason for that they should only be kept in the company of small, active South American Characins, as has been recommended by other authors, if the aquarium is really spacious and the fish can avoid each other when needed. Suitable company fishes are found in the genera *Pterophyllum*, *Mesonauta*, and *Uaru*, with whom the Discus also share their biotopes in nature and which are of comparable disposition. In this connection it needs to be mentioned that Discus fish are very susceptible to infections with protozoan parasites of the genus *Hexamita*. Other Cichlids can be much more resistant to these and may not show any signs of such infection although they are carriers of the parasite. They may thus represent a source of introducing hexamitosis into an aquarium.

Meat of fish and shrimp, cut into appropriate pieces, is an excellent food for adult Discus fish, particularly as freshwater shrimp apparently form a major leg of the diet in nature. Red mosquito-larvae and Tubifex should not be given to these Cichlids as these often originate from highly polluted water and may therefore carry pathogens. The water-temperature of an aquarium for Discus fish should never drop below 27 °C.

There are many beautiful colour varieties of *Symphysodon aequifasciata*

Breeding

For conditioning the fish, the water-temperature should be raised to 29-30 °C. *Symphysodon discus* is an openbrooder which does not spawn on the bottom, but some distance from the ground on vertical surfaces. The brood is reared in a parental family structure.

It is unique among the Cichlids that the juveniles exclusively feed on a skin excrete produced by the parents for the first few days. They die if this food is unavailable. Breeding of Discus fish unfortunately is everything but easy. The most favourable precondition for success is extremely soft water with a pH below 5 that is more or less free of nitrogen-compounds. In order to properly monitor these conditions and to

keep them constant, professional breeders keep their fish in sterile, undecorated tanks with nothing more than an earthenware pot in it that serves as spawning surface. Such keeping of Discus fish must, however, be strictly opposed as it is nowhere near natural. It is also not really a precondition for the successful breeding of these Cichlids.

It is unfortunate that even the presence of eggs is not yet a guarantee for a breeding success as Discus fish often tend to eat their eggs. Even when the larvae have hatched and the fry swims free, their rearing fails simply due to the fact that the parents do not produce the necessary skin excrete.

If everything goes well and the young fish feed on their parents' skin discharge, they should additionally be offered baby Brine shrimp after a few days.

The juveniles of *Symphysodon* species feed on skin excretions of their parents during the first days.

The Genus Theraps
GÜNTHER, 1862

The genus *Theraps* was suggested by GÜN-THER on occasion of the description of *Theraps irregularis*. Subsequently, it was considered to be a synonym of the catch-all genus *Cichlasoma* for decades. REGAN (1905, 1906-1908) later used this name again to create a section within the genus *Cichlasoma* that was a rather heterogeneous assembly of about twenty Central American Cichlids. The division and redefinition of *Cichlasoma* by KULLANDER in 1983 eventually resulted in the necessity that all Central American species be re-allocated to other genera. The genus *Theraps* was one of those cases where a name existed and could be extracted from the former synonymy of the now revised genus *Cichlasoma*.

That the generic name *Theraps* would be available for usage for Central American Cichlids was already pointed out by KULLANDER in 1983. On occasion of the description of two new Mexican Cichlids, SEEGERS & STAECK (1985) and STAWIKOWSKI & WERNER (1987) acted accordingly. ALLGAYER (1989) eventually subjected this genus to a thorough revision as, until then, it was a very heterogeneous assembly of polyphyletic origin. He only accepted eight species of *Theraps,* but several traits used by ALLGAYER to redefine the genus *Theraps* and the allocation of various species were challenged by STAWIKOWSKI & WERNER (1990).

All species of *Theraps* are rheophile fishes exclusive to flowing water bodies with a strong current. Commonly shared traits include amongst others a slender body shape and a small, usually slightly low-set mouth. *Theraps*-species exhibit a great resemblance of the equally rheophile representatives of the genus *Neetroplus*. The most distinctive feature is the shape of the central teeth of the upper jaw that have a

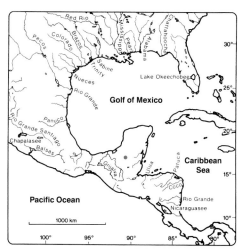

Localities of *Theraps coeruleus*

second apex in the species of *Theraps* but not in the genus *Neetroplus*.

▶ Theraps coeruleus
STAWIKOWSKI & WERNER, 1987

This small Cichlid belongs to a group of rheophile species discovered by French and German aquarists in fast-flowing water bodies in the Mexican state of Chiapas during recent years. Its maximum total length lies close to twelve centimetres.

Theraps coeruleus has a primarily beige coloured body that transforms into a greenish shine towards the belly, particularly in the females. The dorsal and anal fins as well as the ventral fins are pale wine-red, but show a light blue gloss in certain light situations. At the level of the eye, a series of seven, more rarely eight irregularly shaped, black blotches extends from the edge of the gill-cover to the large oval caudal spot. These may expand upwards to reach the base of the dorsal fin in certain moods, thus forming a banded pattern.

Distinguishing between the sexes is easy in the cases of adult specimens as there is a distinct sexual dichromatism present in

Male *Theraps coeruleus*

Theraps coeruleus. Females are characterized by a dark spot on the dorsal fin. Male specimens in contrast have a pattern of tiny black speckles on the flanks and the posterior portion of the gill-cover.

During periods of parental care, both sexes display a very light, slightly blueish to whitish background colouration that is in great contrast with the pattern of black spots and bands. Females have a very distinctive courtship colouration consisting of a blueish green lower head and lower body. Simultaneously, the black spots are replaced by light greenish golden ones.

Natural habitats

Localities of *Theraps coeruleus* recorded to date are exclusive to relatively clear, fast-flowing mountain streams in the drainage of the Rio Mizol Há in the Mexican state of Chiapas. This river flows into the Rio Tulija that in turn forms part of the Rio Grijalva river system. The natural biotopes of *Theraps coeruleus* are characterized by hard, alkaline water.

Care

Theraps coeruleus has turned out to be a robust, adaptable aquarium fish that does particularly well in moderately hard to hard water with an alkaline pH. The decoration of the tank should by all means include a couple of cave-like hiding-places among large rocks and stone flakes as this positively contributes to the comfort of the fish.

Breeding

Theraps coeruleus is a substrate-brooder that preferably spawns inside a cave. The fry is cared for in a father-mother family-structure with a distinct allocation of tasks. Attending to the eggs and larvae is primarily up to the mother fish, while the male parent keeps potential predators out of the breeding territory. The school of young fish is then guided by both parents together. Freshly hatched larvae of the Brine shrimp have proven to be a suitable initial food.

▶ *Theraps nourissati*
ALLGAYER, 1989

Localities of *Theraps nourissati*

This Cichlid was discovered and described only recently. It is the most deep-backed representative of the genus *Theraps*. Some of the specimens caught in the wild, in particular those originating from larger rivers, showed a conspicuous enlargement of the lips, much like in *Amphilophus citrinellus* for example. This hypertrophy is, however, no trait anchored in the genes, but is caused by environmental influences. If these are not provided anymore, for instance in an aquarium, a decrease of the enlargement can be observed.

According to information revealed by the discoverer, specimens from small mountain creeks are the ones with most beautiful colours. The head, particularly its lower portion, shows a deep loam-yellow tone while the gill-covers are metallic blueish green. A wedge-shaped, pale wine-red colour-zone tapers from the nape downwards to the base of the pectoral fin. The rest of the body is brownish yellow to greenish yellow, rather bright yellow in specimens

Theraps nourissati

Female *Theraps nourissati*

engaged in parental care. All unpairy fins are at least partly covered with small greenish glossy speckles.

The intensity of the black lateral pattern is largely influenced by the prevailing mood of a fish. Usually, there is a pattern of seven dark bands between the edge of the gill-cover and the small roundish caudal spot. The anterior two are particularly wide and often look like one double band. At times, the lateral bands may largely be absent, and the fish may then show a wide, irregularly shaped, blackish area on the lower half of the body that almost ranges up to mid-body and then forms a kind of lateral stripe. *Theraps nourissati* may attain a maximum total length of almost twenty centimetres.

Natural habitats

The distribution of *Theraps nourissati* is limited to the state of Chiapas in the south-east of Mexico, and the northwest of Guatemala where the species is endemic to the river system of the Rio Usumacinta.

Locality records exist from the Rio Salinas, the Rio de la Pasión and its tributaries (Rio Puente, Rio Subin), the Rio Lacantum, and the Rio Lacanpo. The species inhabits small clear mountain creeks as well as the large rivers. The biotopes inhabited by *Theraps nourissati* are characterized by moderately hard to hard water and alkaline pH-values.

Care

An appropriate aquarium should contain moderately hard, alkaline water, a bottom covered with sand or fine gravel, numerous hiding-places among stone flakes or respective rock constructions, and be of at least one metre in length.

Breeding

Theraps nourissati is a substrate-brooder which cares for and rears its eggs, larvae, and young fish in a parental family. When the offspring have completed their larval development, they will feed on newly hatched Brine shrimp.

▶ *Theraps rheophilus*
SEEGERS & STAECK, 1985

Localities of *Theraps rheophilus*

A few live specimens of this Cichlid were imported into Europe by several independent parties, including ourselves, for the first time in 1983. The first breeding success followed one year later. The maximum length of this fish is about twelve centimetres.

The background colour is a whitish grey with a pale rosy to reddish tinge on the belly and a yellowish green hue on the back of adult specimens. Seven black vertical lateral bands may appear on the flanks between the edge of the gill-cover and the caudal spot. In addition to this, another two horizontal lateral stripes may be present immediately below the dorsal fin, and at the level of the eye respectively. Where the lateral bands and stripes cross, the pigmentation is particularly dense, so when the bands are suppressed, there seem to be two horizontal rows of irregular black spots. The third lateral blotch is usually particularly large and prominent.

The entire body is covered with tiny black dots. The dorsal fin has a bright red marginal stripe, and except the pectoral fins, all fin-membranes show a pattern of red streaks and dots. At times of parental care, the black markings contrast with a very light, blueish grey background. It is noteworthy that for females engaged in court-

Semi-adult specimen of *Theraps rheophilus*

Theraps rheophilus caring for its brood in the natural habitat

ship, the black pattern vanishes completely, leaving only the third lateral spot that is intensely pale blue then. Male and female specimens are otherwise hard to distinguish on the basis of external traits.

Similar species

Theraps rheophilus was dumped into the synonymy of *Theraps lentiginosus* (STEIN-DACHNER, 1864) by STAWIKOWSKI & WERNER (1987) without justification or reference to the diagnostic traits indicated in the original description. ALLGAYER (1989), in contrast, referred to the specific traits and considered it a good species. *Theraps lentiginosus* indeed has a deeper back, grows much larger (over 20 cm), has a more pointed head, a steeper forehead, smaller eyes, and shorter ventral fins. This species further-more has an additional lateral band, but lacks the Y-shaped forking of the second band, and the pattern of tiny dark dots is not confined to the gill-cover, but extends over the entire lower part of the head up to the lips.

Natural habitats

The only certain locality records come from the Rio Nututun in the Mexican state of Chiapas. This is a small mountain river with a rocky bed filled with cobble. It carries very clear, moderately hard water with dis-tinctly alkaline qualities. According to our observations, this Cichlid is explicitly rheo-phile, which means it preferably resides in those sections of the river where there is a strong current. Sexually active specimens and those engaged in parental care activities were, however, only seen in areas with greatly reduced flow and sandy or muddy bottoms. Other Cichlids in this biotope were *Petenia splendida, Vieja bifasciata, Chuco intermedius, Parapetenia salvini,* and *Thorichthys helleri.*

Care

As this Cichlid is fairly active, it requires quite a lot of space. Its long-term hus-bandry will therefore be only successful in an aquarium of about one and a half metres in length. When the tank is decorated, it should be taken into consideration that larger rocks are arranged to form a number of caves. The water chemistry generally requires little attention. In order to ensure a healthy diet, pieces of fish and shrimp meat should complement the offerings of daphnia and insect larvae.

Breeding

This species is a cavebrooder. In the natural biotope, we could observe that the fish would remove the sand from under flat stones and then use the created cavity for spawning. Both sexes participate in caring for and protecting their young.

Table 37:

Site:	Rio Nututun near Palenque (Chiapas, Mexiko)
Clarity:	ca. 3 m
Colour:	greenish
pH:	8.85
Total hardness:	12.5 °dH
Carbonate hardness:	12.0 °dH
Conductivity:	1200 µS at 27 °C
Depth:	150 – 350 cm
Current:	strong
Water temperature:	27 °C
Air temperature:	30 °C
Date:	2. 4. 1983, 14.00 h

The Genus Thorichthys
MEEK, 1904

was suggested by MEEK in 1904. REGAN (1908) included these Cichlids in the genus *Cichlasoma* and degraded the name to a usage as section. Some subsequent authors then granted it the status of a subgenus, but since the revision and redefinition of the heterogenous generic assembly *Cichlasoma* by KULLANDER (1983), the name *Thorichthys* again enjoys full generic recognition.

The species of *Thorichthys* are very closely allied to one another and there is also a great resemblance between individual species so that a thorough revision is urgently required. MILLER & NELSON (1961) distinguished between eight species, but there are more representatives known from Mexico that have not yet been described. The distribution of *Thorichthys* is limited to the Atlantic slope of Central America covering the area between the drainage of the Rio Antigua in the south of the Mexican state of Veracruz and the region dominated by the Rio Montagua in southeastern Guatemala and northwestern Honduras.

According to MILLER & NELSON (1961), the genus is defined amongst others by the lack of scales on the base of the soft rayed section of the dorsal and anal fins, a cut to convex caudal fin whose outer rays are filamentous in adult specimens, long pectoral fins, a pointed head with a long muzzle, a lateral blotch on the third lateral band, reddish coloured gill-membranes, blue spots on the head and the body, and a black spot on the lower angle of the gill-cover.

The habitus of all *Thorichthys*-species bears a fair resemblance to the Eartheaters of the genus *Geophagus*. As they are relatively small Cichlids reaching maximum lengths of twelve to fifteen centimetres, they are of some importance for the aquarium hobby. The fact that most of the species are very colourful, with shades of red and blueish green glossy spots, only contributes to their popularity.

Some species, such as *Thorichthys meeki*, are robust and adaptable, others, however, have proven to be rather difficult to keep in captivity. They are often very shy and if not provided with an optimal environment, they develop signs of sickness quite rapidly.

White water biotope in the Rio Solimoes with the Giant Water-lily *Victoria amazonica*

▶ *Thorichthys ellioti*
MEEK, 1904

The specimen illustrated here was captured by SEEGERS during a collecting trip to Mexico in 1983 and identified by MILLER. This species can reach a length of approximately fourteen centimetres.

The body of this fish is coloured grey to beige. Six dark bands cover the flanks between the gill-cover and the caudal peduncle with the third band including a distinctly visible lateral blotch. A second black blotch is situated on the lower angle of the gill-cover. The cheeks and gill-covers are marked with relatively expansive blue glossy spots. The majority of scales on the lower half of the body as well as all unpairy fins also carry pale blue glossy spots. Like the dorsal fin, the anal fin has a pale blue submarginal stripe. The chest and belly

Localities of *Thorichthys ellioti*

regions are reddish. The sexes are difficult to distinguish on the basis of external traits.

Thorichthys ellioti

Specific traits

of *Thorichthys ellioti* are the relatively large blue spots on the lower half of the head and a comparatively low dorsal fin. A number of very

similar species

are found in the genus *Thorichthys* so that the identification of this Cichlid is not easy. Useful hints about the identity are provided by the localities: *T. helleri* (STEINDACHNER, 1864) was described from the Rio Teapa (Tabasco) and has a fairly high dorsal fin and smaller blue spots on the head. Other similar species from Mexico are *T. callolepis* (REGAN, 1904) from the upper drainage of the Coatzacoalcos (Oaxaca), and *T. champotonis* (HUBBS, 1936) from the Rio Champoton (Campeche). The

natural habitats

of *T. ellioti* lie in the Mexican state of Veracruz. Locality records centre in the upper drainage of the Rio Tonto which is a tributary to the Rio Papaloapan. The type specimen originates from near Motzorongo. The specimen on the photograph was found in an isolated pool south of Ciudad Aleman, i.e. somewhat below the point where the Rio Tonto flows into the Rio Papaloapan. The water there was very turbid and the bottom muddy. The banks of this water body were partly lined with reed, its surface was partially covered with a carpet of floating plants mainly consisting of *Eichhornia*. Other fishes collected here included Characins *(Astyanax)* and Swordtails *(Xiphophorus)*. The

care

of *Thorichthys*-species is generally easy as they are fairly undemanding and adaptable. The water chemistry is therefore a largely neglectable aspect. Due to their relatively small size, a tank of about one metre in length is spacious enough. Hiding-places

The representatives of the genus *Thorichthys* often live among plants.

are imperative and must be part of the decoration. Plants will not be damaged by these Cichlids.

Breeding

is also not difficult. All species of *Thorichthys* are openbrooders which mostly favour flat stones as spawning surfaces. Both parents very intensely care for and protect their eggs and young so that a few of them will survive even in a community aquarium.

Table 38:

Site:	Stagnant pool south of Ciudad Aleman below of the point where the Rio Tonto flows into the Rio Papaloapan (Vera Cruz, Mexico)
Clarity:	Extremely turbid
Colour:	loam-yellow
pH:	6.75
Total hardness:	5°dH
Carbonate hardness:	5°dH
Conductivity:	850 µS/cm
Depth:	< 1 m
Current:	none
Water temperature:	31°C
Air temperature:	30°C
Date:	29.3.1983, 16.00 h

▶ *Thorichthys meeki*
(BRIND, 1918)

Localities of *Thorichthys meeki*

Thorichthys meeki undoubtedly ranks among the most commonly known Cichlids in the aquarium hobby. For decades it has been a standard fish in the aquarium trade as it is regularly bred by commercial breeders. Reasons for its popularity include a conspicuous colour-pattern, robustness, and adaptability. The maximum total length of the males may come close to fifteen centimetres, and they thus become much larger than the females. The fish mature on reaching a length of approximately ten centimetres. Sexes are difficult to distinguish as the enlarged unpairy fins of fully grown males are the only usable indicator. A junior synonym for this species is *Cichlasoma hyorhynchum* Hubbs, 1935 (comp. HASSE 1981).

Thorichthys meeki has the deep-backed, laterally greatly compressed body typical of the genus. The background colour consists of a shade of light blueish grey. An identification trait is a signal-red coloured zone that extends over the lower portion of the body, from the lower lip to the end of the anal fin, often also including the ventral and anal fins. The presence and intensity of the black markings depend on various moods.

Thorichthys meeki

186

They comprise six broad bands between the edge of the gill-cover and the small caudal spot. The third band is enhanced by a lateral spot on the upper half of the body. This pattern is complemented by a horizontal lateral stripe that runs from the eye to the lateral spot, and an ocellate spot on the lower angle of the gill-cover that is particularly conspicuous due to its silvery halo. Except the uncoloured pectoral fins, all fins carry blueish green glossy dots and streaks. The head and the flanks may also be marked with such glossy dots. The dorsal fin has a narrow signal-red margin.

Natural habitats

The distribution of *Thorichthys meeki* spans over the northeast of Mexico, Belize, and northern Guatemala. Locality records exist amongst others from the river system of the Rio Usumacinta, from Rio Candelaria, Belize River, and water bodies in the vicinity of the city of Progreso in the north of the Yucatán Peninsula. We caught *Thorichthys meeki* in the Mexican states of Tabasco and Chiapas in 1983. Collecting sites were water bodies along the road from Macuspana to Catazaja, including Agua Blanca that all belong to the drainages of the Rio Grijalva and Rio Usumacinta. Here, the fish lived together with *Amphilophus robertsoni* and *Parapetenia octofasciata* in moderately hard, flowing streams with a pH between 7 and 8.5.

Care

A compatible pair can easily be housed in an aquarium of about one metre in length. A community tank in which this species is kept in the company of Central American Cichlids of the genera *Archocentrus* or *Thorichthys* or South American *Aequidens*, however, should measure at least one and a half metres in length. It should be decorated with larger rocks used to structure the bottom space in individual territories. Additional decoration items are pieces of bog-oak among which the fish find places to hide. As this species occasionally nibbles at delicate plants, their usage is limited to hardy species of the genera *Echinodorus* and *Anubias*.

Breeding

Thorichthys meeki is a typical openbrooder that spawns on stones or roots. Both parents participate in caring for the young, but assume certain sex-specific tasks. The immediate care for the eggs and larvae is left to the female fish, while the male primarily defends the breeding territory against potential predators of the offspring. On those occasions, the antagonistic behaviour is very impressive. It consists of spreading the gill-covers and displaying the signal-red gill- and gular skin.

When the young fish begin to swim freely, they are guided by both parents. Initial food offerings should consist of crushed flake-food and newly hatched Brine shrimp.

Table 39:

Site:	Turbid creek a few kilometres south of the village of Macuspana, Rio Tulija drainage (Tabasco, Mexico)
Clarity:	turbid
Colour:	loam-yellow
pH:	7.5
Total hardness:	3 °dH
Carbonate hardness:	4 °dH
Conductivity:	140 µS/cm
Depth:	< 1 m
Current:	minor
Water temperature:	26.5 °C
Air temperature:	32.0 °C
Date:	1. 4. 1983
Time:	11.00 h

The Genus Tomocichla
REGAN, 1908

This genus was suggested by REGAN on occasion of the description of *Tomocichla underwoodi*. It was intended to point out that this Cichlid assumes a special position in the systematics of the Cichlidae due to its unusual posterior location of its ventral fins. MEEK later included this taxon in the genus *Cichlasoma* where it remained for a long time. As a result of the revision of this catch-all genus by KULLANDER (1983), *Tomocichla* was eventually revalidated. At present, it holds only two species.

Localities of *Tomocichla sieboldii*

▶ Tomocichla sieboldii
(KNER & STEINDACHNER, 1864)

This Central American Cichlid was originally allocated to the genus *Heros*. REGAN (1908) then transferred it to the genus *Paraneetroplus*. Synonyms include *Herichthys underwoodi* REGAN, 1905, *Cichlasoma punctatum* MEEK, 1909, and *Theraps terrabae* JORDAN & EVERMANN, 1927. First imports of live specimens into Germany were observed in 1980 thanks to WEBER & WERNER. The maximum length lies near 30 centimetres.

Pair of *Tomocichla sieboldii* during parental care

The fish has a yellowish grey background colouration with a pattern of six to seven black spots or incomplete vertical lateral bands that are usually most prominent at mid-body. A dark spot in the centre of the caudal peduncle is a characteristic feature, and the same applies to two almost horizontally arranged dark lines on the forehead that connect the anterior margins of the eyes. Behind the pectoral fins there generally is a pale reddish to rosy colour-zone. The dorsal fin has a narrow red marginal stripe. Parts of the unpairy fins show a pattern of small brownish red spots.

This species is characterized by a very conspicuous courtship colouration with the black markings becoming particularly obvious. The lower portions of the head and the gill-cover then assume a glossy black colour.

With this species being isomorphic, males and females are very difficult to distinguish although male specimens grow larger. During periods of parental care the females are readily recognized on the basis of the black colouration of the spinous portion of the dorsal fin.

Specific traits

include the rounded caudal fin, the presence of 11-13 soft rays in the dorsal fin, five, in exceptional cases only four, spines, and 8-9, rarely 10, soft rays in the anal fin.

Tomocichla sieboldii is also characterized by the two horizontal black lines on the forehead and the black colouration of the lower part of the head. A very

similar species

is *Tomocichla tuba* (MEEK, 1912). In fact it shares so many traits that both Cichlids are considered sister-species that have their predecessor in common. *T. tuba*, however, has a cut to convex caudal fin, 14-16 soft rays in the dorsal fin, and four, rarely five spines and 10-11 soft rays in the anal fin. Another identification feature is the broad

black mask over the eyes during periods of parental care.

Tomocichla sieboldii only has a relatively small distribution with its

natural habitats

being exclusive to the Pacific slopes of Costa Rica and Panama. The vicinity of Esparta (Rio Jesus Maria, Rio Barranca) appears to be the western limit, whereas the drainage of the Rio San Pedro in Panama represents the eastern border. The fish inhabits a wide variety of biotopes. They live in flowing as well as in stagnant water bodies and appear to be fairly tolerant regarding the hardness and pH of the water. According to BUSSING (1975) this Cichlid is a vegetarian that feeds on algae and grass, as well as leaves and seeds of the terrestrial vegetation. In addition to this, it also takes aquatic insects, their larvae, and freshwater shrimp. The

care

of this Cichlid is not a problem as long as it is housed in a spacious aquarium of one to two metres in length. If it is to share its artificial environment with other Cichlids, the tank even need to be much larger. In order to create hiding-places, to structure the available ground space, and to simplify the definition of territorial borders, large rocks and roots are required. It is not very sensible to try to establish plants in the aquarium. Since the fish need a nutritious diet, they should regularly be offered cut shrimp and freshwater or marine fish. Provided with an adequate environment,

breeding

is actually quite easy. The fish spawn on stones or roots. The eggs, larvae, and young fish are cared for and protected by both parents jointly. When the fry has completed their larval development, they must be fed with the nauplii of Brine shrimp and adequately fine flake-food.

◗ *Tomocichla tuba*
(MEEK, 1912)

Localities of *Tomocichla tuba*

This Central American Cichlid of the section *Theraps* had already been described as *Tomocichla underwoodi* by REGAN in 1908. Subsequently it was decided that there would not be a real justification to have a special genus for this species, and it was therefore included in the section *Theraps* within the genus *Cichlasoma* (BUSSING 1975). As the species name was preoccupied in the genus *Cichlasoma*, MEEK described the Cichlid again as *Cichlasoma tuba* in 1912. SCHULZ was the first to import this Cichlid alive into Germany in 1980 and the first successful breeding in captivity was managed by DOST in 1983. The fish can reach approximately 30 centimetres in length.

The back is greenish grey, the belly rather yellowish. Up to eight or nine, highly irregular lateral bands may be present on the flanks, neither reaching the back nor the belly. Individual scales in the area behind

the pectoral fins carry a red dot. These dots may occasionally even cause the impression of a red lateral stripe that extends up to the caudal peduncle. Small red spots are also present on the posterior portions of the dorsal and anal fins. The iris is equally red. Both sexes display a highly contrasting courtship pattern with the black markings

Tomocichla tuba

on the head and the body standing out from a light background. At these times, a broad black mask becomes obvious spanning from the forehead to the upper lip crossing the eyes.

Tomocichla tuba is polymorphous. Particularly in the areas to the west of its distributions, some specimens have unusually thick lips like those found in some Cichlids from the African Lakes Malawi and Tanganyika or in *Amphilophus labiatus* (GÜNTHER, 1864) from Nicaragua. All intermediate stages, from normal to extremely enlarged lips, may be found at the very same locality. The modification has nothing to do with sex though; males and females are almost impossible to distinguish on the basis of external traits. Only older male specimens develop a humped forehead.

Specific traits

that allow the identification of this species, include the cut to convex caudal fin of adult specimens and the presence of 14-16 soft rays in the dorsal, the four, in exceptional cases five, spines, and 10-11 soft rays in the anal fin. Another characteristic feature is the broad black mask over the eyes during periods of parental care.

A particularly

similar species

is *Tomocichla sieboldii* (KNER & STEINDACHNER, 1863), a species that is in fact so similar that both are considered sister-species originating from the same predecessor. This species, however, has a rounded caudal fin, 11-13 soft rays in the dorsal fin, five, in exceptional cases four, spines, and 8-9, more rarely 10, soft rays in the anal fin. During periods of parental care, there is no mask-like pattern over the eyes, but there are two narrow lines that link the eyes.

The

natural habitats

of *Tomocichla tuba* lie on the Atlantic slope of Central America in the countries Nicaragua, Costa Rica, and Panama. The northern limit of the distribution is formed by the drainage of the Rio Escondido, while the Rio Cricamola represents the southeastern border. The home waters of this Cichlid are usually relatively soft, but generally of a distinct alkaline quality. These fish often inhabit water bodies with a strong current.

According to BUSSING (1975), they are vegetarians feeding mainly on algae as well as leaves and fruit of the terrestrial vegetation. In addition, they prey upon aquatic insects and freshwater shrimp. Even when kept in moderately hard water, the

care

has proven to be easy. As the species is fairly active, it requires a quite spacious aquarium of at least one and a half metres. The tank should be furnished with washed gravel, large rocks, and roots in a way that there are various hiding-places and territories. These fish can easily be kept together with other Central American Cichlids of similar sizes. A healthy and varying diet should include regular offerings of fish and shrimp meat. When *Tomocichla tuba* reaches a length of approximately ten centimetres, the aquarium should be stripped of plants.

Provided a compatible pair has bonded,

breeding

is relatively easy. *Tomocichla tuba* shows a distinct tendency towards cavebrooding. In the first successful breeding attempts the fish spawned inside a large earthenware tube. Both parents participate in caring for and protecting their young. Larvae of the Brine shrimp and finely crushed flake-food are appropriate for initial feedings.

The Genus Uaru
HECKEL, 1840

The 19th century saw descriptions of several species of *Uaru*. However, these were subsequently all considered to be synonymous with *Uaru amphiacanthoides* thus leaving the genus monotypical for a long time. In 1989, STAWIKOWSKI described *Uaru fernandezyepezi* from Rio Atabapo as a new species. In addition, another, taxonomically undescribed member of this genus has been known since the late seventies.

The distribution of the genus *Uaru*, which forms part of the tribe Cichlasomini, ranges over the central Amazon basin and the upper Orinoco and Rio Negro. The genus is defined by its species having 7-10 spines in the anal fin and numerous, very minute scales (often more than fifty in one horizontal row). Close cladistic relationships appear to exist with the genera *Hoplarchus* and *Symphysodon*.

◆ Uaru amphiacanthoides
HECKEL, 1840

belongs to the Cichlids longest kept in captivity. In Germany, for example, the species has been known since 1913. *Uaru obscurum* GÜNTHER, 1862, *U. imperialis* (STEINDACHNER, 1879), and *Pomotis fasciatus* SCHOMBURGK, 1843, are obviously junior synonyms of this species.

With a maximum length in excess of 25 cm, the fish is regularly encountered on the food market in Brazil. Although this Cichlid lacks bright colours, it is extraordinarily conspicuous in its colouration. On a greyish green to greenish yellow background, there is an expansive, elongate, black blotch on the lower half of the body that tapers posteriorly, roughly resembling a wedge. Another black, often rectangular spot is situated on the caudal peduncle. Two more are found on the anterior portion of the body with one lying immediately be-

Localities of *Uaru amphiacanthoides*

hind the eye, and the other, smaller one, above the base of the pectoral fin. The iris of this fish is pretty for its red-orange colouration. Being isomorphic, sexes are difficult to distinguish with certainty on the basis of external characters alone.

Specific traits

that identify *Uaru amphiacanthoides* include the expansive, approximately wedge-shaped, black blotch on the lower half of the body that ends some distance below the upper lateral line, and the two small spots behind the eye.

Similar species

In the late seventies (SCHMETTKAMP 1980), a Cichlid appeared in the aquarium hobby that, at first glance, looked like *Uaru amphiacanthoides*, but in fact has a number of traits that suggest that it be an undescribed species. Its wedge-shaped blotch is more expansive and does not end below, but above the lateral line. The two spots behind the eye are absent. The upper margin of the eye and the clearly visible teeth are red. Finally, this species is more slender and less deep-backed.

The

natural habitats

of *Uaru amphiacanthoides* consist primarily of typical black and white water rivers that are very poor in dissolved minerals and have an acidic pH. The literature contains locality records for the entire Amazon River and the Rio Solimoes, the lower Ucayali, Madeira, Xingú, the Rio Negro, and Rio Branco.

The care

of this Cichlid is usually not coupled to specific problems. It is a peaceful and calm aquarium fish which has rather little space requirements despite its size. Nevertheless

Catching Cichlids in the Amazonian rainforest can become a strenuous business despite purposeful gear.

Uaru amphiacanthoides

should its aquarium be clearly longer than one metre in length. Hiding-places among roots or sensibly arranged stone flakes will greatly contribute to the thriving of this species. Plants should not be used for the decoration of the tank as this Cichlid is a vegetarian. This fact seems to oppose the observation that the natural biotopes of this species are actually bare of aquatic plants. But, during the seasons of high water, the rainforest is submerged for metres and the leaves of the terrestrial vegetation that are now below the water surface represent a source of food for many species of fish. Besides shrimp and fish meat, this fish should therefore also be offered lettuce leaves that, however, must be free of any remains of insecticides.

Breeding

has only been successful in a few isolated cases. It has certainly to do with the fact that the species is so rarely imported and grows so large. It is therefore difficult to assemble compatible breeding pairs. This is complemented by the factor that this Cichlid is apparently quite well adapted to the water chemistry of its natural biotopes. For breeding attempts it is therefore recommendable to make very soft and acidic water available. The fish is an openbrooder which prefers to spawn on vertical surfaces. The brood is reared in a parental family-structure in a manner that shows remarkable parallels with the Discus fish of the genus *Symphysodon* with whom they also share their natural habitats. Like these Cichlids, the young of this species feed on a skin excrete of their parents for the first few days. However, while this form of nourishment is obligatory for baby Discus fish, it is merely an additional source of food for *Uaru amphiacanthoides*. Upon completion of the larval development, the young fish can also be reared on a diet of newly hatched larvae of the Brine shrimp. It is remarkable that this species has an explicit juvenile colour-pattern that consists of a brownish basic colour and a multitude of greenish silvery glossy spots and streaks.

◗ Uaru sp.

This representative of the genus *Uaru* has not yet been dealt with taxonomically, but has been known in the aquarium hobby since the late seventies. To date, imports to Germany were rare and only consisted of single specimens (SCHMETTKAMP 1980, STAWIKOWSKI 1989). The fish has a yellowish brown to greyish green background colouration. The area of the forehead can show a weakly developed pattern of vermiculate brownish red lines. The upper edge of the golden yellow coloured iris is purplish red. The black markings are comprised of a broad lateral band on the caudal peduncle, a spot on the base of the pectoral fin, and an expansive zone on the central and lower flank. The teeth are conspicuous for their brownish red colour. The maximum size of this Cichlid lies at about 30 centimetres.

Specific traits

The most significant colour feature of this Cichlid is the almost rectangular shape of the lateral blotch whose upper margin lies above the upper lateral line. This is complemented by the vermicular pattern on the forehead and the band on the caudal peduncle.

Similar species

In *Uaru amphiacanthoides* the lateral blotch ends a considerable distance below the upper lateral line. The band on the caudal peduncle is replaced by a mere spot, and the eyes are surrounded by a black zone.

Uaru fernandezyepezi has a wide lateral band on the posterior third of the body and a pattern of horizontally arranged small black speckles on the anterior portion. A lateral blotch is absent.

At present, there is no reliable information on the origin of this undescribed Cichlid which was found in shipments from Manaus and is offered as edible fish on the markets of this city. It can only be assumed that it occurs in the drainage system of the central Amazon River.

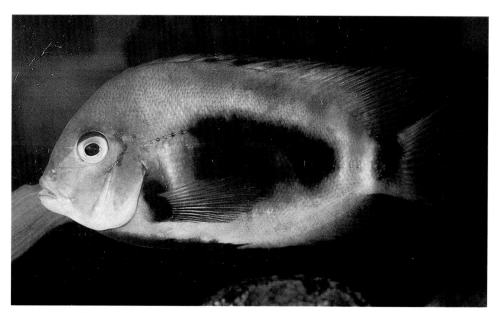

Uaru sp., a species yet to be dealt with taxonomically

The Genus Vieja
FERNANDEZ-YEPEZ, 1969

For decades, *Vieja* was considered a synonym of the genus *Theraps*. It was only after the splitting of the genus *Cichlasoma* by KULLANDER (1983) and the revision of *Theraps* by ALLGAYER (1989) that it became valid again. A junior synonym is *Paratheraps* WERNER & STAWIKOWSKI, 1988 (ALLGAYER 1991). At present, *Vieja* contains about 10 species.

White water biotope of *Uaru amphiacanthoides* in the Rio Solimoes near Manaus. The fish have access to ample vegetarian food at times of high tide.

◗ *Vieja bifasciata*
(STEINDACHNER, 1864)

Localities of *Vieja bifasciata*

This Central American Cichlid was originally described as *Heros bifasciatus* and previously assigned to the section *Theraps* of *Cichlasoma*. Individual live specimens were first imported into Germany by SCHULZ in the late seventies. More imports were due to ourselves as well as STAWIKOWSKI & WERNER in 1983.

This species may attain a length of about 30 centimetres. The most significant traits in the colour-pattern of this species are two black horizontal stripes on the flanks. The lower stripe begins above the base of the pectoral fin on the edge of the gill cover and stretches up to the base of the caudal fin. The upper stripe is approximately situated in the area of the upper lateral line. It begins slightly more posteriorly than its lower counterpart and ends below the base of the last rays of dorsal fin or at the anterior portion of the caudal peduncle. This stripe may not always be clearly defined, but can also appear as a vague black colour-zone. The head of this fish, mostly also including its nape and the anterior region of the back, shows an intense green to greenish yellow sheen. Throat, chest, and the belly area are red, while the rest of the body is intensely golden yellow. Adult specimens exhibit small red spots on the head and nape. The caudal fin ends with a broad red marginal

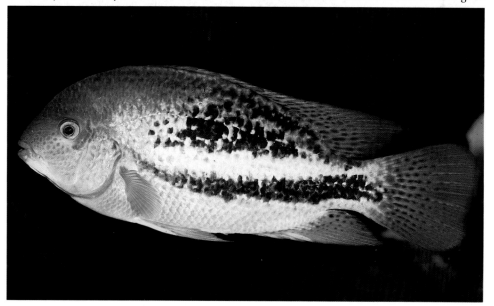

Vieja bifasciata

band. The soft rayed parts of the dorsal and anal fins are also reddish.

This species is isochromatic and isomorphic, meaning there are no differences in colour or shape that would allow an identification of the sexes.

Specific traits

of *Vieja bifasciata* primarily include the colour-pattern that lacks dark vertical bands, but shows two black horizontal stripes. Moreover, the red colouration of the throat, chest, and belly regions is characteristic. The genus *Vieja* holds several very

similar species,

among which *Vieja guttulata* (GÜNTHER, 1864) and *Vieja fenestrata* (GÜNTHER, 1860) must be mentioned. Both these species have more or less the same body shape and partly even comparable colour-patterns, but both lack the upper horizontal stripe that is so typical of *Vieja bifasciata*.

The

natural habitats

of this Cichlid lie in the northwest of Guatemala (vicinity of the Laguna de Petén) and in the Mexican state of Chiapas (drainage of the Rio Usumacinta). ALVAREZ (1970) also mentioned the river system of the Rio Coatzacoalcos, but this was certainly based on an error. We observed and caught this species in the Rio Nututun, in the vicinity of the town of Palenque. This is a typical clear water river with a rocky bottom and a considerable current. Here, *Vieja bifasciata* preferably resided in the deeper, calmer sections of the river where there was sufficient cover among submerged branches or large rocks. Pairs engaged in parental care were, however, mostly seen in shallow sections with sandy bottom. An analysis of the water revealed a distinctly alkaline pH. Other Cichlids in this biotope were *Petenia splendida*, *Chuco intermedius*, *Parapetenia salvini*, *Theraps rheophilus*, and *Thorichthys*

Vieja bifasciata in its natural habitat

We cought *Vieja bifasciata* in typical clear water biotopes with a strong current.

helleri. The habitat was furthermore shared with the Characin *Astyanax fasciatus,* the Live-bearing toothcarp *Poecilia sphenops,* and *Chirostoma* sp., Catfishes, and Sleepers.

The long-term

care

of this large Cichlid necessitates a very large aquarium of more than one and a half metres in length. It should be decorated with large rocks and roots that are arranged in such a manner that some hiding-places are created. You should largely forgo planting the tank. It is only recommendable to choose a few large, robust individual plants, e.g. species of *Anubias* or ferns that anchor themselves on the rock constructions. In order to achieve optimum colouration of the fish, the diet need to be given particular attention. A varying feeding schedule should therefore include pieces of fish and shrimp meat. Provided with a compatible pair,

breeding

does not cause problems. *Vieja bifasciata* is a typical openbrooder which prefers flat stones as spawning media. Both parental specimens participate in caring for and protecting the offspring. Baby Brine shrimp and pulverized flake-food have proven suitable as initial food for the young fish.

Table 40:

Site:	Rio Nututun near Palenque (Chiapas, Mexico)
Clarity:	ca. 3 m
Colour:	greenish
pH:	8.85
Total hardness:	12.5 °dH
Carbonate hardness:	12.0 °dH
Conductivity:	1200 µS at 27 °C
Depth:	150 – 350 cm
Current:	strong
Water temperature:	27 °C
Air temperature:	30 °C
Date:	2. 4. 1983
Time:	14.00 h

◆ *Vieja fenestrata*
(GÜNTHER, 1860)

This Central American Cichlid was temporarily assigned to the section *Theraps* within *Cichlasoma*. Originally, it was described as *Chromis fenestrata*. STAWIKOWSKI & WERNER were the first to import live specimens into Germany in 1983. It may grow to a length of more than 20 centimetres.

The fish has a greyish green to greenish body colouration with the back showing a metallic golden shine. Head, nape, throat, and chest are bright wine-red in adult specimens. This colouration is also present on the posterior portions of the dorsal and caudal fins. The belly, and the ventral and anal fins, are mainly black. A black lateral stripe runs over the flank from the edge of the gill-cover to the centre of the caudal peduncle. Between the pectoral fin and the base of the caudal fin, this band is crossed by six or seven black lateral bands. Slightly above the edge of the gill-cover, the nape

Localities of *Vieja fenestrata*

often shows a conspicuous black blotch representing the remains of another lateral band. Sexes are isomorphic which means they cannot be identified with certainty on the basis of external traits. Male specimens often exhibit more red in their fins though.

Vieja fenestrata

This species obviously is polychromatic as, according to our observations, there are almost uniformly reddish specimens in Lake Catemaco. However, these are very rare.

Specific traits

useful for the identification of this species, are such as the pattern of six to seven dark vertical bands crossing a lateral stripe, and the red colouration of the head. This is complemented by the presence of more than 32 scales in a horizontal row and only seven or eight gill-rakers on the lower branch of the first gill-arch.

The genus *Vieja* includes several

similar species

which have more or less identical body-shapes, a broad red margin of the caudal fin, and black stripes or bands on the flanks. These include *Vieja maculicauda* (REGAN, 1905), *Vieja guttulata* (GÜNTHER, 1864), and *Vieja hartwegi* (TAYLOR & MILLER, 1980). None of these Cichlids has a red head though, and the pattern of lateral bands and stripes is usually not complete in these species.

The

natural habitats

of *Vieja fenestrata* primarily centre in water bodies on the Atlantic side of the Mexican states of Veracruz and Oaxaca. The distribution ranges from the drainage of the Rio Chachalacas, and the Rio Actopan north of the city of Veracruz, through the drainage of the Rio Papaloapan, up to the Rio Coatzacoalcos. Most authors are of the opinion that this species would not occur east of this river, but ALVAREZ (1970) published records from the Rio Champotón and Rio Usumacinta. This is confirmed by the data supplied by ALMAZAN & MEDINA (1974) who found this Cichlid in the vicinity of the Laguna de Terminos, in the border region between the states of Tabasco and Campeche. We encountered this species in Lake Catemaco near the town of San Andres Tuxtla. Due to its wide distribution, this Cichlid inhabits quite a variety of water bodies which, in their majority, have a neutral to slightly alkaline pH and that are soft to moderately hard. This results in the

care

not being particularly difficult. Accounting for its size, this Cichlid requires quite a lot of space so that only tanks with a length clearly greater than one metre are suitable. Large rocks and roots should be arranged to create hiding-places in order to provide comfort for the fish. If plants are to be used, these should be large and hardy. Provided with a large enough aquarium, *Vieja fenestrata* can be kept in the company of other Central American Cichlids. The relatively large food requirements can be complied with by regular offerings of pieces of lean fish and shrimp meat. With a compatible pair being a precondition,

breeding

hardly poses a problem. Conditioning necessitates quantitatively sufficient feedings and a good quality of the water. First signs of an intention to breed are courtship displays that become more and more frequent with time. These include jerking the head and vibration of the body. At a later stage, the fish dig in the bottom substrate and clean objects that seem suitable for spawning. Shortly before mating, they undertake to guide and lure the partner to the spawning site. Spawning takes place in the fashion of openbrooders whereby a stone is usually preferred. The female as well as the male participate in caring for and protecting the eggs, larvae, and young fish. Feeding them with *Artemia salina* and crushed flake-food, rearing is effortless.

▶ *Vieja hartwegi*
(TAYLOR & MILLER, 1980)

This Central American Cichlid was originally assigned to the section *Theraps* of *Cichlasoma* and belongs to those species that
were discovered and described only
recently. First imports into Europe were to
the credit of STAWIKOWSKI & WERNER in
1983, with the first successful captive breeding following in the subsequent year. The
largest known specimen in the literature has
a length of about 15 centimetres.

The background colouration of the fish
varies with the mood, ranging from a lighter
to a darker brownish yellow overlain by a
beautiful, mostly blueish green metallic
gloss. The flanks are marked with a dark

Localities of *Vieja hartwegi*

lateral stripe that only begins some distance
from the gill-cover and extends up to the

Breeding *Vieja hartwegi*

caudal spot. A series of five dark blotches is, however, often shown by the fish instead of an uninterrupted stripe. These may be vertically enlarged to form bands, but never reach onto the back. The forehead is usually marked with two narrow lines that link the anterior edges of the eyes. The head and the anterior portion of the body are covered with a multitude of small wine-red spots and dots. The hind region of the caudal fin, and the margin and soft part of the dorsal fin are red. Sexually active specimens often display a blackish lower head, throat, chest, and belly. The sexes are difficult to identify on the basis of external traits, but old males have larger fins. The

specific traits

that identify *Vieja hartwegi* include the slightly protruding or overhanging upper jaw, and the pattern of several bands and a stripe, or a horizontal series of blotches respectively, that only begin shortly before mid-body.

Several closely related and

similar species

are found in the genus *Vieja*. These are such as *Vieja bifasciata* (STEINDACHNER, 1864) with its two lateral stripes and no bands, *Vieja fenestrata* (GÜNTHER, 1860) with its red head, and *Vieja guttulata* (GÜNTHER, 1864) with a distinct spotted pattern on the head. *Vieja hartwegi* has no wide distribution. Its

natural habitats

are exclusive to the Mexican state of Chiapas, i.e. to the drainage of the upper Rio Grijalva (Rio Grande de Chiapa, Rio Pando, Rio Frio, Rio Salado). Published locality records centre in the wider vicinity of the towns of Villa Flores, Tuxtla Gutierrez, and Comitán. The home waters of this Cichlid are usually moderately to very hard with a pH around 7.5 (STAWIKOWSKI 1983).

Further information was provided by TAYLOR & MILLER (1980), according to which the fish are not adapted to a certain bottom substrate and can be found above sand and mud as well as above gravel, cobble, or rocks. Except species of *Potamogeton*, there was no aquatic vegetation in these biotopes. Fishes from the same localities were amongst others the Cichlid *Parapetenia grammodes*, Killifishes of the genus *Profundulus*, the Catfish *Rhamdia guatemalensis*, a Live-bearing Toothcarp (*Poecilia sphenops*), and Characins representing the genera *Astyanax* and *Brycon*.

Generally speaking, the

care

of *Vieja hartwegi* does not pose any particular problems. The aquarium should have a minimum length of about one metre. It is decorated with washed gravel or sand, larger rocks and roots, and a few robust plants in such a manner that there are hiding-places and the structure of the ground-space makes the definition of territorial borders easy. The water chemistry does not require close monitoring in the case of this species. In order to provide the fish with a varied diet, they should be offered *Daphnia*, White and Black mosquito larvae, and pieces of fish and shrimp in alternation.

Following such regime makes

breeding

easy. Like all species of *Vieja, Vieja hartwegi* is an openbrooder that spawns on rocks or roots. The eggs, larvae, and young fish are cared for by both the female and the male. Rearing the young on a diet of the larvae of *Artemia salina* and crushed flake-food is generally simple. Attention needs to be given to regular partial exchanges of the water in order to prevent harmful metabolism products from reaching dangerous levels in the aquarium.

► *Vieja maculicauda*
(REGAN, 1905)

There are a number of synonyms for this Cichlid, including *Cichlasoma globosum*, *C. manana*, and *C. nigritum* that were all suggested by MEEK in 1907. This species belongs to the small group of Cichlids that were kept in Germany as early as before the First World War. Subsequently it became extinct in the hobby and was forgotten. It was then not before the late seventies that live specimens were imported again. The fish may reach a length of approximately 30 centimetres.

Localities of *Vieja maculicauda*

Vieja maculicauda is a Cichlid with splendid colours. Its grey to greyish green background colour shows some delicate metallic sheen. A black blotch is present between the ninth and the twelfth ray of the dorsal fin. It continues in the form of a vertical band of the same colour on the flanks. Another black spot lies in the central area of the base of the caudal fin. The lower portion of the head and the throat are vermilion in colour while the forehead between the upper lip and the eyes shows a shade of brownish yellow. The ventral, dorsal, and anal fins are blueish green and have a pattern of black speckles. In contrast, the caudal fin is bright red. The splendid colours of

Vieja maculicauda

the fish fade slightly during periods of parental care and are then dominated by slate-grey tones. The dark iris simultaneously becomes lemon yellow.

With a sexual dimorphism being absent, the sexes are difficult to distinguish. Male specimens, however, grow distinctly larger and develop larger fins later in their lives.

Several geographical colour-morphs of *Vieja maculicauda* are known which primarily differ with regard to the extent of the red and black colour-zones. One population of unknown origin is present in the aquarium hobby that is conspicuous for its body sides being covered with small black dots. The most significant

specific traits

that identify this Cichlid are the black body-belt and the red caudal fin.

Similar species

are found in the genus *Vieja* that also accommodates *Vieja maculicauda*. *Vieja bifasciata* and *Vieja melanura*, for example, have more or less the same body shape and a comparable colouration. In these two species, however, the black markings on the flanks are not arranged vertically, but horizontally, i.e. parallel to the longitudinal axis of the body.

Natural habitats

Vieja maculicauda inhabits a vast area in Central America, ranging from the Rio Chagres in Panama, through Cost Rica, the Atlantic side of Nicaragua (Lake Nicaragua), up to Belize (Golden Stream) and Guatemala (Lake Izabal). PERRONE (1978) published interesting ecological information obtained from a study conducted in the biotopes of Panamanian populations. According to this, reproduction activities are initiated by the male which occupies a territory of one to two metres in diameter and courts every female ready to spawn that happens to come past. The fish do not have

Cichlid biotope near Tamasopo (San Luis Potosi, Mexico)

a restricted reproduction period, but spawn throughout the year. In their natural habitats, they also live in the brackish water of the lower parts of the rivers. It would therefore be unsuitable to keep them in water poor in minerals or at a distinctly acidic pH.

The care

of this robust and adaptable Cichlid is easy if its size is accounted for and it is housed in a very spacious tank of about one and a half metres in length. The availability of cave-like hiding-places among larger rocks or pieces of bog-oak obviously contributes to the thriving of the fish. Adult specimens require a nutritious diet and should regularly be offered pieces of fish and shrimp meat.

Breeding

requires a compatible pair as a precondition, but is otherwise simple. The fish usually spawn on a flat rock. At a water-temperature of approximately 28 °C, the larvae hatch after 48 hours. The young start swimming around after five days. In nature, the young are cared for by the parents for about 40 to 50 days, until they have reached a length of about 22 millimetres.

◗ *Vieja* sp. cf. *guttulata*
(El Aguacero)

Localities of *Vieja* sp. cf. *guttulata*

According to MILLER (in litt.), this Central American Cichlid of the genus *Vieja* has not yet been scientifically described. The specimen figured here was caught by us in 1983. The maximum length of this species should lie at approximately 25 centimetres.

The most significant component in the pattern of this Cichlid is a black lateral stripe that stretches from the upper angle of the gill-cover to the centre of the caudal peduncle, but does not extend onto the caudal fin. This is complemented by five or six vertical bands that neither reach the back nor the belly and whose presence or absence depend on the mood. The background colour of the fish is greyish green to brownish yellow. The lower head has a delicate greenish to greenish golden-metallic sheen that is overlain with a fairly conspicuous pattern of small reddish brown to red spots. This spotted pattern also extend onto the nape, the anterior back, and the chest.

Vieja sp. cf. *guttulata*

The posterior edge of the caudal fin as well as the outer portion of the soft part of the anal and dorsal fins are vividly red in colour. External sexual traits are not known.

Specific traits

in the colour-pattern of this Cichlid that are indicative include the pattern of small reddish spots on the head and anterior body, and the arrangement of vertical black lateral bands and one horizontal stripe on the flanks. The genus *Vieja* holds several fairly

similar species

of which *Vieja guttulata* (GÜNTHER, 1864), described from Guatemala, deserves particular mentioning as it appears to be closely related. This Cichlid is said to have the pattern of lateral bands restricted to the upper half of the body (REGAN 1905). The

natural habitat

of this yet to be described Cichlid lies in the Mexican state of Chiapas. The locality is situated on the El Aguacera waterfall of the Rio Flores. This river flows into the Rio de la Venta and is thus connected to the Rio Grijalva. The water at this site was extraordinarily clear and had a fairly strong current. The bottom consisted of fine brownish sand with occasional larger limestone outcrops. The water of the Rio Flores is moderately hard with distinct alkaline qualities. Other fishes found at this location, were the Killifish *Profundulus labialis*, Live-bearing Toothcarps, and Characins. The

care

of this Cichlid has turned out to be simple. Its long-term husbandry obviously requires spacious tanks of about one and a half metres in length. This Cichlid also needs a few cave-like hiding-places among large rocks and roots to feel comfortable. The usage of plants can only be recommended

El Aguacero, biotope of *Vieja* sp. cf. *guttulata*

with reservations. Keeping this fish in captivity does not require a close monitoring of the water chemistry. A nutritious and varying diet is of great significance and should include regular offerings of pieces of meat of shrimp and freshwater and marine fish. Up to when this manuscript was completed,

breeding

had not yet been successful. It is supposed that its breeding ecology does not differ significantly from other species of *Vieja*, i.e. that it is an openbrooder and rears its offspring in a parental family, with both parents jointly protecting the fry from predators.

Table 41:

Site:	Waterfall El Aguacero (Rio Flores drainage, Chiapas, Mexico)
Clarity:	clear
Colour:	none
pH:	8.8
Total hardness:	9°dH
Carbonate hardness:	10°dH
Conductivity:	2800 µS at 26.5°C
Depth:	< 1 m
Current:	strong
Water temperature:	26.5°C
Air temperature:	30.0°C
Date:	5.4.1983, 14.00 h

◗ *Vieja synspilum*
(HUBBS, 1935)

A synonym of this Central American Cichlid of the genus *Vieja* is *Cichlasoma hicklingi* FOWLER, 1956. First live imports into Europe took place towards the end of the seventies. Male specimens may easily reach 30 centimetres in length.

Fully coloured specimens certainly rank among the most splendid Cichlids we know. As the colouration is, however, so variable, it is difficult to provide a credible description. In general, it is simple to recognize and differentiate individual specimens on the basis of personalized peculiarities. The largest portion of the body usually shows an attractive shade of blueish green. The posterior half often has expansive golden yellow or orange coloured zones. These are frequently covered with small blackish spots and streaks that also extend onto the belly and back in some specimens. The throat and chest, but also sometimes the entire head and the nape are bright red. In most the

Localities of *Vieja synspilum*

cases the cheeks and forehead are yellowish to orange though. Some individuals display a marvellous blueish green colouration around the eye. The iris is also blueish green. Approximately five black blotches are present on the posterior half of the body that often fuse to a highly irregular lateral stripe. As there are no distinct differences between males and females, a reliable ident-

Vieja synspilum

ification of the sexes based on external characteristics is an impossible task. Although older, almost fully grown males develop a hump on the forehead as is so typical of many Cichlids, and thus the upper profile is almost vertical this trait is, however, usually of little usefulness as it develops very late. The most indicative

specific trait

that makes it easy to identify this species is the colour-pattern. The red head is a characteristic feature and so is the black lateral band or series of blotches that are confined to the posterior half of the body. The humped forehead of old specimens is also typical of this species. Although the genus *Vieja* contains several

similar species,

confusion is unlikely to take place as only one more species has a red head. This is *Vieja fenestrata* (GÜNTHER, 1860) which, on the other hand, has a characteristic pattern of one black lateral stripe and several bands. The

natural habitats

of *Vieja synspilum* lie on the Atlantic slope of Central America. The distribution ranges from the Mexican states of Tabasco (Rio Usumacinta, Rio Tabasquillo) and Campeche (Rio Champoton), through the north of Guatemala to Belize, the former British Honduras (Belize River, Rio San Pedro). The water bodies inhabited by this species are characterized by soft to moderately hard water with a pH that may either be slightly acidic or slightly alkaline. This already suggests that this Cichlid be fairly adaptable and its

care

in captivity has proved this to be correct. Although adult specimens are very quiet

This vegetated biotope is also the home of Cichlids (Laguna Media Luna, Mexico)

fish with no explicit urge to move, their successful long-term keeping definitely requires a very spacious aquarium of more than one and a half metres in length. Hiding-places among large slabs of rock or pieces of bog-oak contribute to the fish feeling safe. Planting the aquarium is less recommendable and may only be sensible if a few large, hardy, individual plants are used. The fish require a rich diet and should therefore regularly be offered meat of fish and shrimp (*Mysis*). Ox-heart should only be given occasionally as regular offerings of this type of food causes the beautiful colours of the fish to fade.

Breeding

is not always successful in the first attempt. It is fundamental that the fish accept each other as partners. They are openbrooders that attach the eggs to rocks or roots. Both parental specimens care for and protect the brood.

With larger females laying several hundred eggs at a time, regular partial exchanges of the aquarium water are necessary to ensure that the metabolism end-products do not poison the water. The young fish are easily reared on a diet of baby Brine shrimp and crushed flake-food.

◗ *Vieja zonata*
(MEEK, 1905)

According to MILLER (1966) *Vieja zonata* is only a synonym as he is of the opinion that MEEK's description refers to a species from Guatemala that had already been described as *Heros guttulatus* by GÜNTHER in 1864. This opinion was also shared by some subsequent authors. However, we are using this name in a valid sense as the fish figured here were caught in the vicinity of the type locality of *Vieja zonata* and differ in some aspects from *Vieja guttulata*. STAWIKOWSKI & WERNER were the first to import this species into Europe in 1983. The length of the fish supposedly exceeds 25 centimetres. Temporarily it was assigned to the section *Theraps* within *Cichlasoma*.

Adult specimens have an attractive greenish golden metallic sheen on the upper half of the body. In contrast, throat, chest, and belly are blueish. The ventral and anal

Localities of *Vieja zonata*

fins show bright shades of blue. The outer portion of the caudal fin, and the soft part of the dorsal fin, are red. Below the lower branch of the lateral line, the body appears to be mingled by numerous small spots that may densify to form five or six vertical

Vieja zonata

210

bands. Juveniles exhibit six or seven lateral bands as well as one lateral stripe. Sexes cannot be identified with certainty on the basis of external characters alone.

Specific traits

Features of the colour-pattern of *Vieja zonata* that allow an identification of the species, are the mainly greenish blue background colours, and the black horizontal lateral stripe that extends from the hind edge of the eye up to the base of the caudal fin in both juveniles and adults. It is furthermore significant that there is no distinct spotted pattern on the lower part of the head and no well-defined vertical bands on the sides of the body.

Similar species

The fairly variable clade centring around *Vieja guttulata* (GÜNTHER, 1864), a species described from Lake Amatitlan in Guatemala, is particularly difficult to distinguish from *Vieja zonata*. One colour trait of that species, however, is the presence of a well-defined pattern of often reddish spots that extends up to the lower portion of the head. Furthermore, *Vieja guttulata* has a more pointed head and a less deep back than *Vieja zonata*. Finally, that species shows several vertical bands on the upper half of the body.

Natural habitats

The localities of *Vieja zonata* lie in the Mexican state of Oaxaca. Its distribution appears to be restricted to rivers flowing into the Pacific Ocean. The type locality is situated in the vicinity of the village of Niltepec, about 50 km east of the town of Juchitan. The natural biotopes of this species carry relatively soft water with a pH ranging between 6.5 and 7.5 (STAWIKOWSKI 1983).

Care

Keeping this Cichlid in an aquarium has turned out to be simple. The long-term husbandry necessitates spacious tanks with lengths clearly over one metre, though. The aquarium should be decorated with washed gravel or sand, larger rocks, and roots. Some hiding-places contribute to the fish feeling safe. Planting can only be recommended with the reservation that only robust, hardy plants are used.

Due to their size the fish require a nutritious diet that must be ensured by feeding them with cut meat of shrimp and freshwater and marine fish.

Breeding

does not dictate particular conditioning. *Vieja zonata* is a typical openbrooder which prefers flat stones as spawning media. During periods of parental care, the female and male form a parental family-structure to care for the offspring and defend them against approaching predators. Initial feedings for the young fish may consist of baby Brine shrimp and crushed flake-food.

Table 42:

Site:	River in the vicinity of La Ventosa, about 40 km west of Niltepec (Estado de Oaxaca, Mexico)
Clarity:	relatively turbid
Colour:	brownish
pH:	7.75
Total hardness:	15°dH
Carbonate hardness:	11°dH
Conductivity:	1150 µS/cm
Depth:	< 150 cm
Current:	swift
Water temperature:	30°C
Air temperature:	34°C
Date:	6.4.1983
Time:	11.00 h

211

INDEX OF SYNONYMS

Obsolete name	Presently valid name
Acarichthys geayi	= *Guianacara geayi*
Aequidens geayi	= *Guianacara geayi*
Cichlasoma alfari	= *Amphilophus alfari*
Cichlasoma bartoni	= *Parapetenia bartoni*
Cichlasoma bifasciatum	= *Vieja bifasciata*
Cichlasoma bulleri	= *Paraneetroplus bulleri*
Cichlasoma carpinte	= *Herichthys carpintis*
Cichlasoma citrinellum	= *Amphilophus citrinellus*
Cichlasoma cyanoguttatum	= *Herichthys cyanoguttatus*
Cichlasoma dovii	= *Parapetenia dovii*
Cichlasoma ellioti	= *Thorichthys ellioti*
Cichlasoma fenestratum	= *Vieja fenestrata*
Cichlasoma festae	= *Parapetenia festae*
Cichlasoma grammodes	= *Parapetenia grammodes*
Cichlasoma guttulatum	= *Vieja guttulata*
Cichlasoma hartwegi	= *Vieja hartwegi*
Cichlasoma intermedium	= *Chuco intermedius*
Cichlasoma labiatum	= *Amphilophus labiatus*
Cichlasoma labridens	= *Parapetenia labridens*
Cichlasoma maculicauda	= *Vieja maculicauda*
Cichlasoma managuense	= *Parapetenia managuensis*
Cichlasoma meeki	= *Thorichthys meeki*
Cichlasoma nicaraguense	= *Copora nicaraguensis*
Cichlasoma panamense	= *Neetroplus panamensis*
Cichlasoma sajica	= *Archocentrus sajica*
Cichlasoma septemfasciatum	= *Archocentrus septemfasciatus*
Cichlasoma sieboldii	= *Tomocichla sieboldii*
Cichlasoma spilurum	= *Archocentrus spilurus*
Cichlasoma synspilum	= *Vieja synspilum*
Cichlasoma temporale	= *Hypselecara temporalis*
Cichlasoma trimaculatum	= *Parapetenia trimaculata*
Cichlasoma tuba	= *Tomocichla tuba*
Cichlasoma urophthalmum	= *Parapetenia urophthalma*
Cichlasoma zonatum	= *Vieja zonata*
Geophagus acuticeps	= *Satanoperca acuticeps*
Geophagus daemon	= *Satanoperca daemon*

BIBLIOGRAPHY

ALLGAYER, Robert (1988): Redescription du genre *Paraneetroplus* REGAN, 1905, et description d'une espèce nouvelle du Mexique (Pisces, Perciformes, Cichlidae). Rev. fr. Cichl., *75:* 4—22.

— — (1989): Révision et redescription du genre *Theraps* GÜNTHER, 1862: Description de deux espèces nouvelles du Mexique (Pisces, Perciformes, Cichlidae). Rev. fr. Cichl., *90 bis:* 4—30.

— — (1991): *Vieja argentea* (Pisces, Teleostei, Cichlidae) Une espèce nouvelle d'Amérique centrale. Rev. fr. Cichl., *114:* 2—15.

— — (1994): Description d'une espèce nouvelle du genre *Archocentrus* GILL & BRANSFORD, 1877 (Pisces, Cichlidae) du Panama. Rev. fr. Cichl., *135:* 6—24.

ALMAZAN, Silvia Toral & A. R. MEDINA (1974): Los cichlidos (Pisces: Perciformes) de la Laguna de Terminos y sus afluente. Rev. Biol. Trop., *21* (2): 259—279.

ALVAREZ DE VILLAR, J. (1970): Peces mexicanos (claves). Ser. invest. pesq., estudio No. 1, Inst. nac. invest. biol. pesq., Mexiko.

ARTIGAS-AZAS, J. M. (1993): *Herichthys tamasopoensis* ARTIGAS-AZAS, 1993, ein neuer Cichlide aus Mexiko. Cichlidenjb., *3:* 65—70.

ASTORQUI, I. (1971): Peces de la cuenca de los grandes lagos de Nicaragua. Rev. Biol. Trop. *19* (1,2): 7—57.

BARLOW, George W., et al. (1976): Chemical Analyses of Some Crater Lakes in Relation to Adjacent Lake Nicaragua. In: Investigations of the Ichthyofauna of Nicaraguan Lakes. ed. Thomas B. THORSON. Univ. Nebrasca, Lincoln: 17—20.

— — & J. W. MUNSEY (1976): The Red Devil-Midas-Arrow Cichlid Species Complex in Nicaragua, in: THORSON, T. B. (ed.): Investigations of the Ichthyofauna of Nicaraguan Lakes. Univ. Nebraska, Lincoln: 359—369.

BARAN, Gerd (1981): Die Zucht von *Acarichthys heckelii*. DCG-Info, *12* (10): 185—190.

BAYLIS, Jeffrey R. (1974): The Behavior and Ecology of *Herotitapia multispinosa* (Teleostei, Cichlidae). Z. Tierpsychol., *34:* 115—146.

BURGESS, W. E. (1981): New information on the Species of the Genus *Symphysodon* with the Description of a New Subspecies of *S. discus* HECKEL. TFH, *29* (3): 32—42.

BUSSING, W. A. (1966): New Species and New Records of Costa Rican Freshwater Fishes with a Tentative List of Species. Rev. Biol. Trop. *14* (2): 205—249.

— — (1974): Two new species of Cichlid fishes, *Cichlasoma sajica* and *C. diquis*, from southeastern Costa Rica. Rev. Biol. Trop., *22* (1): 29—49.

— — (1975): Taxonomy and biological aspects of Central American cichlid fishes *Cichlasoma sieboldii* and *C. tuba*. Rev. Biol. Trop., *23* (2): 189—211.

— — (1987): Peces de las Aguas continentales de Costa Rica. San José, Costa Rica. 271 pp.

— — & M. MARTIN (1981): Systematic status, variation and distribution of four Middle American cichlid fishes belonging to the *Amphilophus* species group. genus *Cichlasoma*. Contrib. Sci. (269): 1—41.

CICHOCKI, F. (1976): Cladistic History of Cichlid Fishes and Reproductive Strategies of the American Genera *Acarichthys, Biotodoma* and *Geophagus*. 2 vols., Univ. Michigan, Ph. D. 710 pp.

ECKINGER, Detlef (1987): Nachzucht von „ *Geophagus*" daemon. DCG-Info, *18* (7): 132—134.

EIGENMANN, C. M. (1910): Catalogue of the fresh-water fishes of tropical and south temperate America. Repts Princeton Univ. Exped. Patagonia 1896—1899. Zool. 3: 375—512.

— — (1912): The freshwater fishes of British Guiana, including a study of the ecological grouping of species and the relation of the fauna of the plateau to that of the lowlands. Mem. Carneg. Mus. 5: 578 pp.

— — (1922): The fishes of western South America. Part I. The freshwater fishes of north-western South America, including Colombia, Panama, and the Pacific slopes of Ecuador and Peru, together with an appendix upon the fishes of the Rio Meta in Colombia. Mem. Carneg. Mus. 9: 1—346.

FOWLER, H. W. (1944): Los peces del Perú. Catálogo sistemático de los peces que habitan en aguas peruanas. Boln. Mus. Hist. nat. „Javier Prado" 8: 260—290.

— — (1954): Os peixes de agua doce do Brasil. Volume II. Archos. Zool. S. Paulo, 9: 1—400.

GOSSE, J. P. (1976): Révision du genre *Geophagus*. Mém. Acad. r. Sci. Outre-Mer, Cl. Sci. nat. méd. N. S. *19* (3): 1—172.

— — & S. O. KULLANDER (1981): The Zoological Name of the Red-Hump *Geophagus* (Teleostei: Cichlidae). Buntb. Bull., J. Am. Cichl. Assn., *No. 83:* 12—17.

GÜNTHER, A. (1866): On the Fishes of Panama. Proc. Zool. London, *39:* 600—603.

— — (1869): An Account of the Fishes of the States of Central America. Trans. Zool. Soc. London, *6:* 377—494.

HASEMAN, J. D. (1911): An annotated catalogue of the cichlid fishes collected by the Expedition of the Carnegie Museum to Central South America, 1907—10. Ann. Carneg. Mus. 7: 329—373.

HASSE, J. H. (1981): Characters, Synonymy and Distribution of the Middle American Cichlid Fish *Cichlasoma meeki*. Copeia (1): 210—212.

HUBBS, Carl L., (1936): Fishes of the Yucatan Peninsula. Carnegie Inst. Wash. Publ., *No. 457:* 157—287.

KÄHSBAUER, P. (1968): Beiträge zur Kenntnis einiger Cichliden (Pisces) von Centralamerika. Ann. naturh. Mus., Wien *72:* 161—175.

KUHLMANN, Friedrich (1984): *Biotodoma* erfolgreich im Aquarium nachgezogen. DATZ, *37* (1): 14—17.

KULLANDER, S.O. (1977): *Papiliochromis* gen. n., a new genus of South American cichlid fish. Zool. Scr. *6:* 253—254.

— — (1980): A taxonomical study of the genus *Apistogramma* REGAN, with a revision of Brazilian and Peruvian species. Bonn. Zool. Monogr. *14:* 152 pp.

— — (1981): A Cichlid from Patagonia. Buntb. Bull., J. Am. Cichl. Assn. *No. 85:* 13—23.

— — (1981): Cichlid fishes from the La Plata basin. Part I. Collections from Paraguay in the Muséum d'Histoire naturelle de Genève. Rev. suisse Zool. *88:* 675—692.

— — (1982): Cichlid fishes from the La Plata basin. Part III. The *Crenicichla lepidota* species group. Revue suisse Zool. *89:* 627—661.

— — (1983): Cichlid fishes from the La Plata basin. Part IV. Review of the *Apistogramma* species, with description of a new species. Zool. Scr. *11:* 307—313.

— — (1983): A Revision of the South American Cichlid Genus *Cichlasoma* (Teleostei: Cichlidae). Swedish Mus. Nat. Hist. Stockholm, 296 pp.

— — (1986): Cichlid fishes of the Amazon River drainage of Peru. Stockholm, 431 pp.

— — & H. NIJSSEN (1989): The Cichlids of Surinam: Teleostei, Labroidei. Brill, Leiden, 256 pp.

— — & M. C. SILFVERGRIP (1991): Review of the South American cichlid genus *Mesonauta* GÜNTHER (Teleostei, Cichlidae) with descriptions of two new species. Rev. suisse Zool., *98* (2): 407—448.

LINKE, H., & W. STAECK (1992): Amerikanische Cichliden I: Kleine Buntbarsche. Tetra-Verlag, Melle, 232 pp.

— — (1994): American Cichlids I: Dwarf Cichlids. Tetra Sales, Blacksburg (USA) & Eastleigh (GB). 232 pp.

LOISELLE, Paul (1982): Our National Cichlid, *Cichlasoma cyanoguttatum* Baird & Girard 1854. FAMA, *5*(5): 6ff.

LOPEZ, M. J. (1974): Variación, coloración y estado sistimatico del pez centroamericano *Cichlasoma nicaraguense* (familia Cichlidae). Rev. Biol. Trop. *22*(1): 161−185.

LOWE (McConnel), R. H. (1964): The fishes of the Rupununi savanna district of British Guiana, South America. Part 1. Ecological groupings of fish species and effects of the seasonal cycle on the fish. J. Linn. Soc. Lond. Zool. *45*: 103−144.

−− (1969): The cichlid fishes of Guyana, South America, with notes on their ecology and breeding behaviour. Zool. J. Linn. Soc. *48*: 255−302.

LÜLING, K. H. (1973): Die Laguna de Vegueta an der Küste Mittelperus und ihre Fische, insbesondere *Aequidens rivulatus* (GÜNTHER, 1859). Zool. Beitr., N. F., *19*(1): 93−108.

MACHADO-ALLISON, Antonio (1971). Contribución al Conocimiento de la taxonomía del género *Cichla* (Perciformes: Cichlidae) en Venezuela. Pt. I. Acta biol. Venz., *7*: 459−497.

−− (1971). Contribución al Conocimiento de la taxonomía del género *Cichla* en Venezuela. Pt. II: Osteologia comparada. Acta biol. Venez., *8*: 155−205.

MEEK, Seth E. (1904): The Fresh-water Fishes of Mexiko North of the Isthmus of Tehuantepec. Publ. Field Columb. Mus., zool. ser., *5*: 1−252.

MILLER, R. R. (1965): Geographical distribution of Central American freshwater fishes. Copeia, 1966: 773−802.

−− (1976): Geographical distribution of Central American freshwater fishes with Addendum. In T. B. THORSON (ed.) Investigations of the ichthyofauna of Nicaraguan lakes, pp. 125−156.

−− & B. C. NELSON (1961): Variation, life colors, and ecology of *Cichlasoma callolepis*, a cichlid fish from southern Mexiko, with a discussion of the *Thorichthys* species group. Occ. pap. Mus. Zool. Univ. Mich. (622): 1−9.

PERRONE, Michael (1978): Mate size and breeding success in a monogamous cichlid fish. Env. Biol. Fish, *3*(2): 193−201.

PETERS, H. M. & BERNS (1978): Über die Vorgeschichte der maulbrütenden Cichliden. I. Was uns die larvalen Haftorgane lehren. II. Zwei Typen von Maulbrütern. Aqu. Magazin *12*: 211−217, 324−331.

−− (1982): Die Maulbrutpflege der Cichliden: Untersuchungen zur Evolution eines Verhaltensmusters. Z. f. zool. Systematik u. Evolutionsforschung 20(1): 18−52.

OHM, Dietrich (1978): Sexualdimorphismus, Polygamie und Geschlechtswechsel bei *Crenicara punctulata* GÜNTHER, 1863. Sitzungsber. Ges. Naturf. Fr. Berlin (N. F.), *18*: 90−104.

PLOEG, A. (1991): Revision of the South American cichlid genus *Crenicichla* HECKEL, 1840, with descriptions of fifteen species and considerations on species groups, phylogeny and biogeography. Academisch Proefschrift, Univ. Amsterdam. 153 pp.

REGAN, C. T. (1905): A revision of the fishes of the South-american cichlid genera *Crenacara*, *Batrachops* and *Crenicichla*. Proc. zool. Soc. Lond. 1905: 152−168.

−− (1905): A revision of the fishes of the South-American cichlid genera *Acara*, *Nannacara*, *Acaropsis* and *Astronotus*. Ann. Mag. nat. Hist. (7) *15*: 329−347.

−− (1905): A revision of the fishes of the American cichlid genus *Cichlosoma* and of the allied genera. Ann. Mag. nat. Hist. (7) *16*: 60−77, 225−243, 316−340, 433−445.

−− (1906): A revision of the South-American cichlid genera *Retroculus*, *Geophagus*, *Heterogramma* and *Biotoecus*. Ann. Mag. nat. Hist. (7) *17*: 49−66.

−− (1906): A revision of the fishes of the South-American cichlid genera *Cichla*, *Chaetobranchus* and *Chaetobranchopsis*, with notes on the genera of American Cichlidae. Ann. Mag. nat. Hist. (7) *17*: 230−239.

−− (1906−1908): Pisces. Biologia Centrali-Americana, *8*: 1−203.

REIS, R. E., & L. R. MALABARBA (1988): Revision of the Neotropical cichlid genus *Gymnogeophagus* RIBEIRO, 1918, with descriptions of two new species (Pisces, Perciformes). Revta. Bras. Zool., *4*(4): 259−305.

ROGERS, W. (1981): Taxonomic Status of Cichlid Fishes of the Central American Genus *Neetroplus*. Copeia (2): 286−296.

SCHMETTKAMP, W. (1980): Die Gattung *Uaru* HECKEL, 1840. DCG-Info *11*(6): 105−114.

SCHULTZ, L.P. (1960): A review of the Pompadour or Discus fishes, genus *Symphysodon* of South America. TFH, *8*(10): 5−17.

SEEGERS, L., & W. STAECK (1985): *Theraps rheophilus* nov. sp., ein ungewöhnlicher Cichlide Mexikos aus der *Cichlasoma*-Verwandtschaft. DATZ, *38*(11): 499−505.

STAECK, Wolfgang (1982): Segelflosser und Flaggenbuntbarsche − vor Ort beobachtet: Neue Erkenntnisse über ihre natürlichen Lebensräume. Aqu. Magazin, *16* (5): 290−293.

−− (1982): Handbuch der Cichlidenkunde. Franckh. Stuttgart, 200 pp.

−− (1983): Cichliden: Entdeckungen und Neuimporte. Engelbert-Pfriem-Verlag, Wuppertal, 351 pp.

−− (1984): Ein natürliches „Aquarium": Beobachtungen am Rio Nututun in Mexiko. TI, *No. 65* (März): 5−7.

−− & I. SCHINDLER (1993): Anmerkungen zur Gattung *Mesonauta* GÜNTHER, 1862. DCG-Info, *24*(3): 57−68.

−− & Lothar SEEGERS (1984): Die Fische der Laguna Media Luna und der Laguna Los Anteojitos, Rio Verde, Mexiko: 3. Die Cichliden. DATZ, *37*(6): 204−209.

STAWIKOWSKI, Rainer (1983/84): Auf der Suche nach Majorras: Zum Cichlidenfang nach Mexiko, Teil 1−7. DATZ, *36* (8): 289−293; (9): 323−327; (10): 364−368; (11): 411−415; (12): 449−452; *37*(1): 9−13; (3): 87−91.

−− (1989): Ein neuer Cichlide aus dem oberen Orinoco-Einzug: *Uaru fernandezyepezi* n. sp. (Pisces: Perciformes: Cichlidae). Bonn. zool. Beitr. *40*(1): 19−26.

−− (1993): *Mesonauta*: Weitere Beobachtungen und ergänzende Anmerkungen. DCG-Info, *24*(8): 161−174.

−− (1994): Maulbrütende *Heros*! DATZ, *47*(4): 212.

−− & U. WERNER (1985): Die Buntbarsche der Neuen Welt: Mittelamerika. Reimar Hobbing, Essen. 271 pp.

−− & U. WERNER (1987): Corrienteros: Die Gattung *Paraneetroplus* REGAN, 1905. DCG-Info, *18*(11): 209−219.

−− (1987): Neue Erkenntnisse über die Buntbarsche um *Theraps lentiginosus* mit der Beschreibung von *Theraps coeruleus* sp. nov. DATZ, *40*(11): 499−504.

−− (1988): Die Buntbarsche der Neuen Welt: Südamerika. Reimar-Hobbing-Verlag, Essen. 288 pp.

TAYLOR, J.N. & R.R. MILLER (1980): Two new cichlid fishes, genus *Cichlasoma*, from Chiapas, Mexico. Occ. Pap. Mus. Zool. Univ. Mich. (693): 1−16.

−− (1983): Cichlid Fishes (Genus *Cichlasoma*) of the Rio Panuco Basin, Eastern Mexico, With Description of a New Species. Occ. Pap. Mus. Nat. Hist. (Univ. Kansas, Lawrence), *No. 104*: 1−24.

THORSON, Thomas B., ed. (1976): Investigations of the Ichthyofauna of Nicaraguan Lakes. Univ. Nebraska, Lincoln. 663 pp.

TIMMS, A. M. & KEENLEYSIDE, M. H. A. (1975): The reproductive behaviour of *Aequidens paraguayensis*. Z. Tierpsychol. *39*: 8−23.

VILLA, J. (1976): Systematic status of the cichlid fishes *Cichlasoma dorsatum*, *C. granadense* and *C. nigritum* MEEK. In T. B. THORSON (ed.) Investigations of the ichthyofauna of Nicaraguan lakes. pp. 375−383.

WEBBER, R. G., W. BARLOW & A. H. BRUSH (1973): Pigments of a Color Polymorphism in a Cichlid Fish. Comp. Biochem. Physiol. *44B*: 1127−1135.

WERNER, Uwe, & R. STAWIKOWSKI (1990). Anmerkungen zu ALLGAYERS Revisionen der Gattungen *Paraneetroplus* und *Theraps*. DATZ, *43*(7): 442−443.

PHOTO CREDITS

All photographs by Wolfgang Staeck

except those on pages

106, 178, 179, 180 (Jens Gottwald),

30, 72, 102, 132, 133 (Ingemann Hansen),

42 (Lutz Krahnefeld),

32, 38, 44, 54, 58, 67, 73, 84, 86, 87, 97 (Horst Linke),

195 (Erwin Schraml),

184, 206 (Lothar Seegers),

9, 21, 22, 204 (Ernst Sosna),

95, 96 (Frank Warzel),

141, 150, 200, 202 (Uwe Werner).

Drawings: H. Linke

Maps: G. Springer

The Authors

Horst Linke, born in 1938, has had an interest in the aquarium since early childhood. Already quite early the dream of all enthusiastic aquarists to visit the tropical habitats of our aquarium-fishes came true for him.

In 1963, he undertook a journey throughout Black Africa, and two years later he had opportunity to visit the countries of Panama, Venezuela, Peru, and Bolivia. From the contacts with the natural habitats in the home countries of the aquarium-fishes in the wild, new questions and tasks always developed so that he visited some countries not only once but repeatedly.

Beginning in 1973, he undertook collecting and study expeditions to Cameroon, Nigeria, Ghana, Togo, Sierra Leone, Tanzania, Kenya, Thailand, Sumatra, Borneo, Malaysia, Colombia, Peru, and Bolivia in quick succession. During his numerous stays abroad it was always of special interest for him to collect as much information as possible about the life-conditions in the natural biotopes in order to create an optimal environment for the fishes in the aquarium at home.

Over the years, his journeys were planned with more and more precisely defined tasks and specific study-goals, may it be to verify doubtful distribution records or to collect material for the work on taxonomical problems.

He made other aquarists profiting from his experiences by lectures, but especially by publications in both national and international periodicals.

Wolfgang Staeck, born 1939, studied biology and English literature at the Free University Berlin, Germany. After his State Diploma he worked as an associate researcher at the Technical University Berlin for several years. In the year 1972 he conferred a degree with the minor subjects Zoology and Botany.

Dr. Staeck is known to a wide public through numerous lectures and the publication of books and papers in journals. Since 1966 he has been publishing a vast number of contributions on Cichlids in German and foreign magazines. As his major interest is focused on behavioural studies in Cichlids, he is still an aquarist today and familiar with the maintenance and breeding of Cichlids with the experience of many years.

During his numerous study-trips which were primarily intended to learn more about Cichlids in their natural environments and resulted in the discoveries of many new species, subspecies, and colour-varieties, he travelled East Africa especially, but also West Africa and Madagascar. In recent times he undertook journeys to Central and South America to study and collect Cichlids in Mexico, Brazil, Ecuador, Venezuela, Guyana, Argentina, Peru, and Bolivia.

A high priority of his research was spent on the Cichlids of the Lakes Malawi and Tanganyika which he had travelled to already in the early seventies. Not only in these waters but also in rivers of Central and South America, he observed and took photographs of the world of fishes as a diver. Through this he managed to document the ecology and the inhabited biotopes of many Cichlids for the very first time in underwater photographs. As a result of his study-trips he published scientific descriptions of several new species of Cichlids.